Martin Colognoli

Coral

H E M E R I A

Preface

My first Natural History book, *Corals of Australia*, was published in 1986: 644 pages long, all in colour. By then I had published a life-changing five-volume technical monograph about corals. Very satisfying, but something was missing – those volumes said nothing about the beauty of coral reefs, which is what drew me to them in the first place. After *Corals of Australia* was published and my work in Australia was done, I moved north to Indonesia, the Philippines and Japan, always working with local people, always diving from their boats: 'tinnys', canoes with outriggers, and sometimes sailboats. Of course, these people, who depended on coral reefs for their livelihoods, began to fascinate me: they spent their entire lives on them, and by them, fishing, diving, and talking about them. Millions of people.

It wasn't always fun. In most of my years diving in Indonesia and the Philippines I felt I was in a warzone for I would hear explosions, sometimes kilometers away sounding like ricocheting bullets, other times frightenly close. The locals were making improvised bombs from beer bottles filled with garden fertilizer mixed with diesel, complete with newspaper wicks. Skillfully thrown, the explosions smashed the coral and killed or stunned everything under them including fish of all sizes, which these people hastily collected. I became determined to do something about this, for these fishermen were destroying their own livelihoods as well as the reefs they lived by. There were no marine nature reserves in these countries then; now they are collectively known as the Coral Triangle, home to the greatest diversity of life on our planet and a magnet for scientists and tourists alike. Following the greatest conservation effort in maritime history, these reefs have become a source of income for these people; exploitation has turned to conservation. So, my life as a marine biologist – a mixture of science, beauty, caring and people – turned me into a conservationist, and how could it not.

Need I say, I loved this book at first glance. Words, being the vehicle of science, reach out, but only go so far. Photographs illustrating the beauty of reefs go to the heart as well as the head, but again, only so far. The association of words, photographs of reefs and photographs portraying the lives of the people who live by them are a rare combination. My ardent wish is that this book goes far and wide, carrying its message, touching the minds and hearts of people who care, and raising awareness among all those that it reaches.

Charlie Véron

Charlie Veron, known as the "godfather of coral", was born in Sydney in 1945. He is an Australian biologist specialised in the study of coral and their reefs. Veron has named about 20% of reef corals and he discovered and delineated the Coral Triangle. He is also an environmental activist.

Preface

Humans and coral... So different and yet so close. If on the face of it there's no physical resemblance, we do have common roots that have shaped our genetic make-up. Despite evolving separately over 600 million years, 50% of coral's genome can also be found in humans, making coral a perfect model for biomedical research.

Humans and coral... So different, and yet coral is crucial to humanity. Coral reefs teem with biodiversity, home to more animal species per unit of area than large primary forests. An important fact is that 25% of all marine life can be found concentrated in the 0.2% of the ocean's surface made up of coral reefs. They are the largest animal constructions in the world, protecting hundreds of countries' coasts from the effects of waves and erosion. Coral reefs are also a source of subsistence. Reefs are rich fishing grounds and generate a lot of revenue in tourism. Five hundred million people across the globe depend directly on coral reefs for their livelihoods.

Humans and coral... A deep and ancestral relationship, a symbol of purity. For ancient Polynesians, a tiny fragment of coral taken from the reef symbolised a prayer, an *ubu*, a request to the gods to rid the country of its sins so that it would be as pure as the just-harvested coral.

Humans and coral... So close and so intimate that they have almost a symbiotic relationship... But symbiosis is reciprocal. Although coral is essential to mankind, humans are coral's biggest threat. The rise in ocean temperatures has had a dramatic effect on coral, leading to the "white death", or coral bleaching. According to the Global Coral Reef Monitoring Network's (GCRMN) most recent report, 14% of reef-building corals disappeared between 2009 and 2018. For a symbiotic relationship between people and coral to develop, humans must protect coral so that coral reefs don't end up being the first ecosystems to be completely eradicated from the world because of human activity.

With this magnificent collection of photos, Martin Colognoli seeks to move his readers and challenges them to do their part to help coral reefs to survive. I sincerely hope that this book achieves that goal and opens his readers' eyes

Denis Allemand

Denis Allemand is a university professor of animal biology and the scientific director of the Monaco Scientific Centre. His research focuses on the physiology of marine organisms, primarily coral, and their use as model organisms to help understand major theories in biology (such as biomineralization, symbiosis and aging).

Introduction

CORAL tells the story of the rebirth of an Indonesian fishing village. The village's fishermen became protagonists in the restoration of the coral reefs that surround their village, and are now responsible for the conservation of these resources.

CORAL is a story. It's a story of realisation: realising the stunning beauty that surrounds us in nature and our lives. Above all, it's a story about the inextricable bond between humans and living things. As a marine biologist, Martin Colognoli had the opportunity to explore coral ecosystems, and was driven to do something in the fact of an impending, human-made catastrophe: the destruction of coral reefs and the wild species so closely linked to them.

He voluntarily took a step back from science in order to devote himself to coral conservation. It was as a photographer that he was able to transform his passion into art. His photographs, along with Coral Guardian, the NGO that he founded, play an important role in raising awareness about marine conservation.

As well as telling his personal story, Martin Colognoli's overriding wish was to demonstrate solidarity and respect and show how we are an integral part of the animal kingdom. Living far from urban environments where it is all too easy to forget our relationship with nature, he took a particular interest in the way of life of traditional communities who are confronted with the modern world. For these communities, what is the impact of new technologies, new ways of producing, consuming, communicating – always more?

Through his photography, Martin Colognoli pays homage to the village residents with whom he forged relationships over time. All of the photographs are the end result of his experience on the ground over the course of many years, and are dedicated to all of the people who support and inspire his work. His photos capture spontaneous interactions, without any staging or direction. This book comes from his desire to share those moments, mixing what is imagined and real, his beliefs along with innovation. The book also tells us that living in harmony with nature doesn't have to be a way of life that will inevitably disappear.

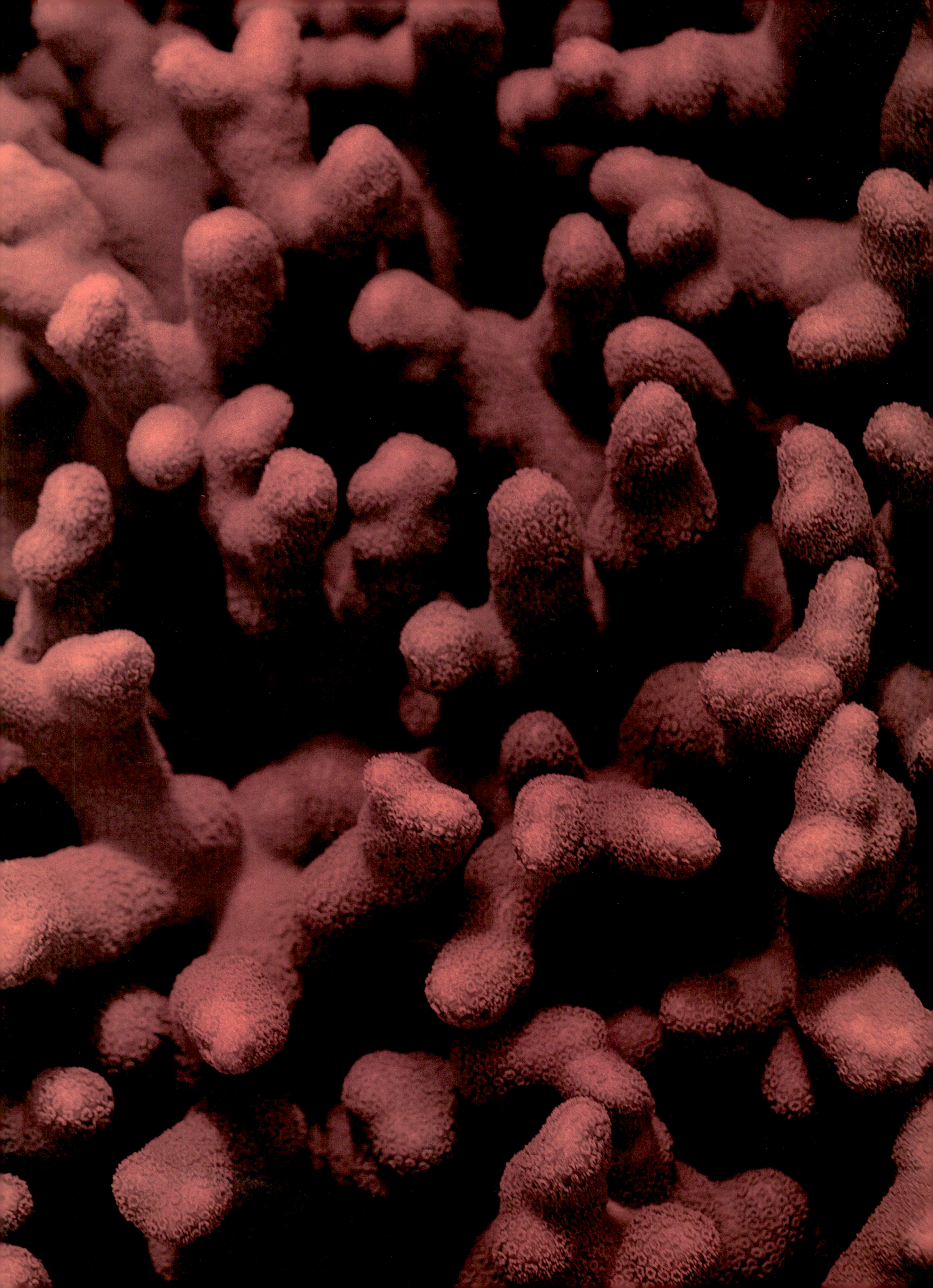

CORAL

At sea level, mysterious underwater beings form the largest structures in the world, ancestral monuments. Complex, fragile and unique systems, simultaneously mineral and vegetable but functioning like an animal. Coral fascinated Martin Colognoli and led the scientist-by-training on an artistic journey. He decided to point his lens at this enigmatic creature in a humble attempt to capture its beauty, strength and delicate nature.

He takes us on a travel to a rock in the middle of the ocean known as Bird Island, Pulau Burung in Indonesian, Tukobele in dialect, or perhaps to Hatamin Island, a strip of land just a few dozen metres long, covered in plant life and enveloped in a 50-hectare coral reef that formed around an ancient volcano emerging from the earth's crust. In these waters, just a few nautical miles separate coral brimming with life from grey deserts of coral skeletons.

The photographer gets as close as possible to the coral, capturing them in black and white to bring out their rich texture, their architectural ingenuity. He contemplates them like an impressionist painter, portraying their diverse shapes, their colours that change in the flickering light, catching the movement of the waves and the fleeting moments when the coral rises to the water's surface.

He shows this animal, so strangely similar to humans, which lives alone or in communities, fighting for territory while all the while adapting to having to share it. He captures its details, seemingly endowed with a sort of collective consciousness, here resembling a human brain, there a lung. He pays homage to this intriguing miracle of nature which, by blooming in water lacking in nutrients, enables other animal and plant life to become abundant.

Despite coral's promise of eternity, its 500 million years of existence, despite the power of the reefs, despite being the only animal construction visible from space, Martin exposes its fragility. Along with ours. Suddenly, face to face with something that can never be repaired, these images seem to forget movement, lose colour like scenes of war. Mineral landscapes of coral bleached by heat, ravaged by dynamite fishing or destroyed by a stray anchor.

And what will happen to humans?

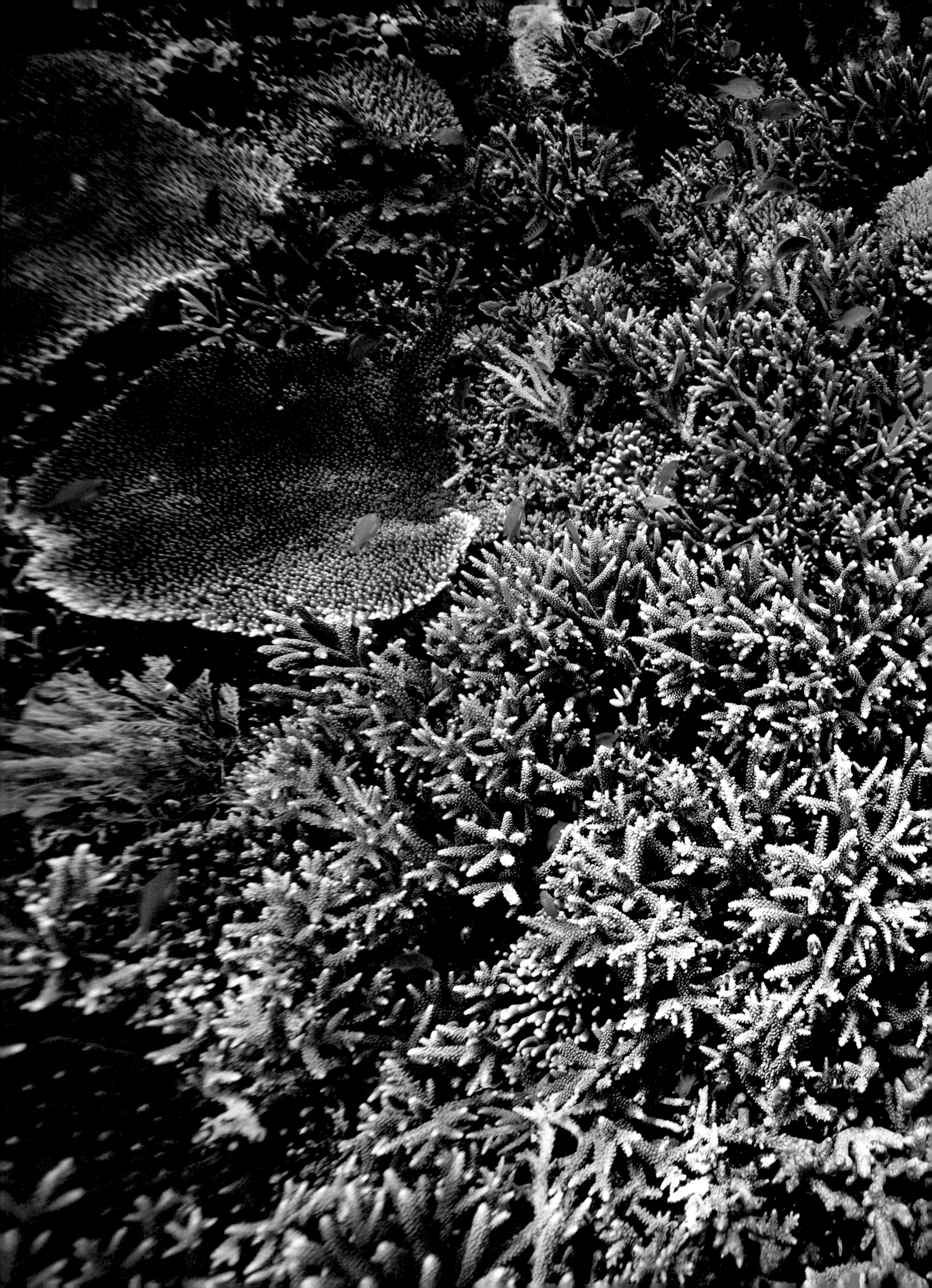

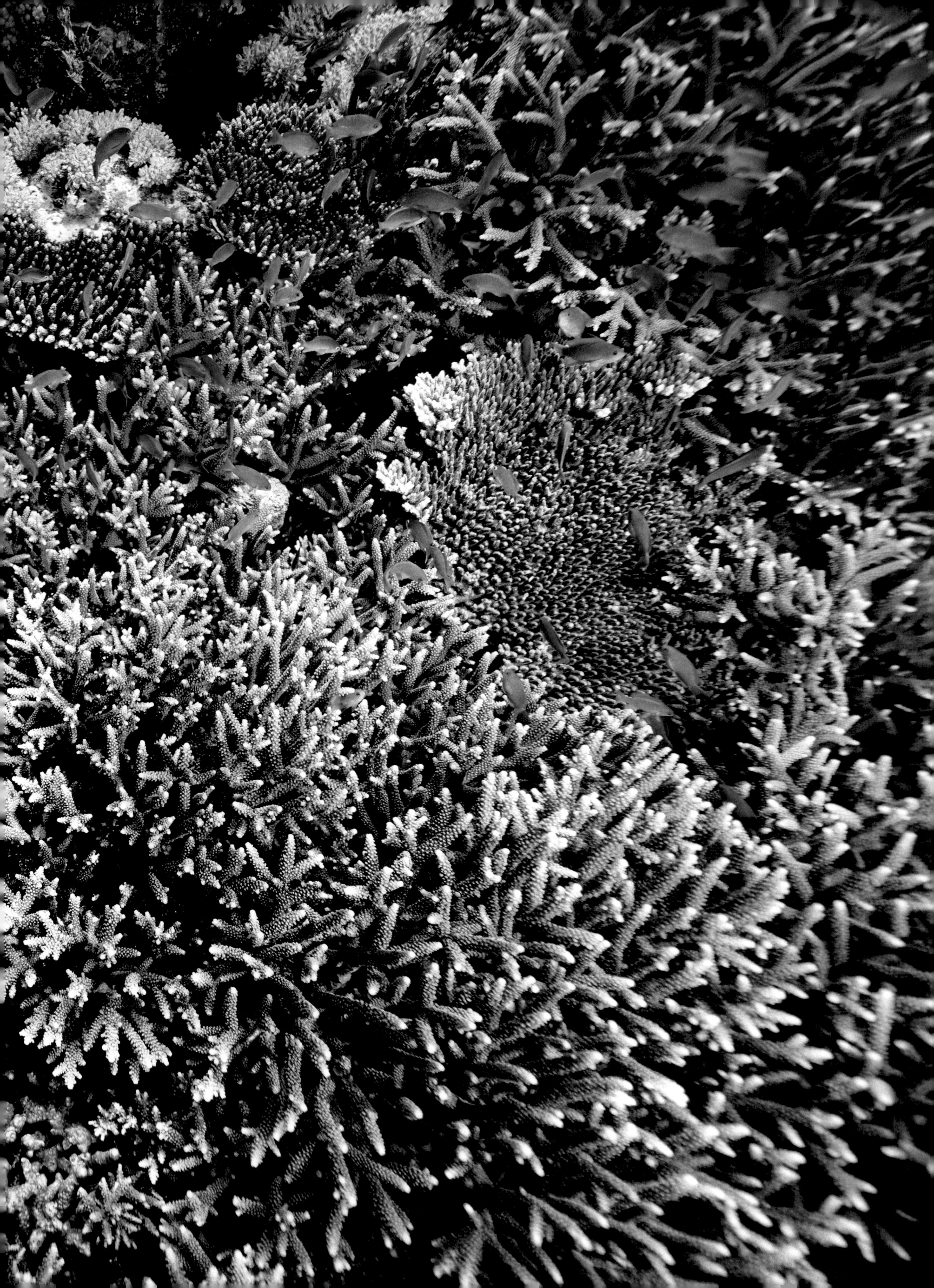

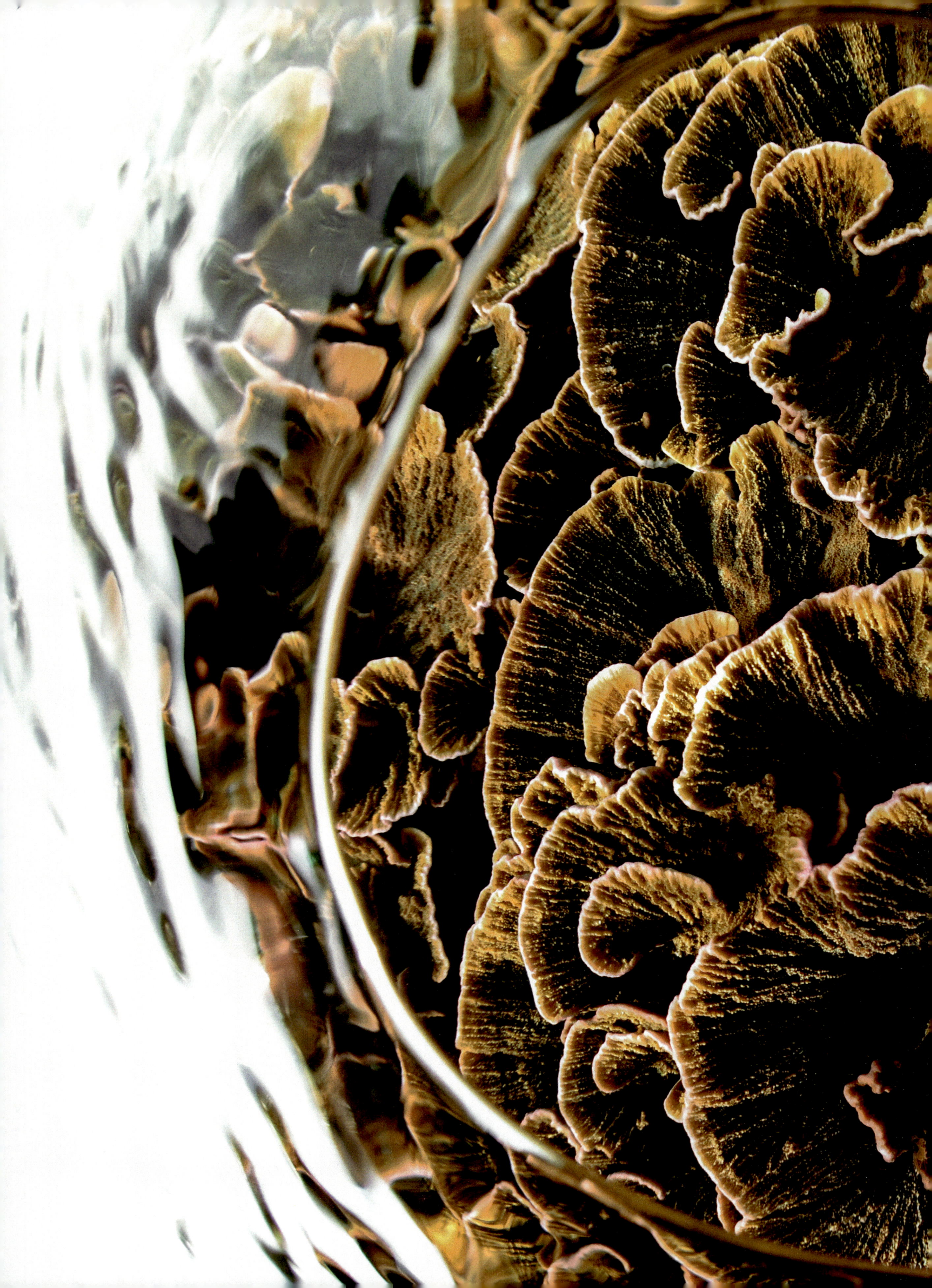

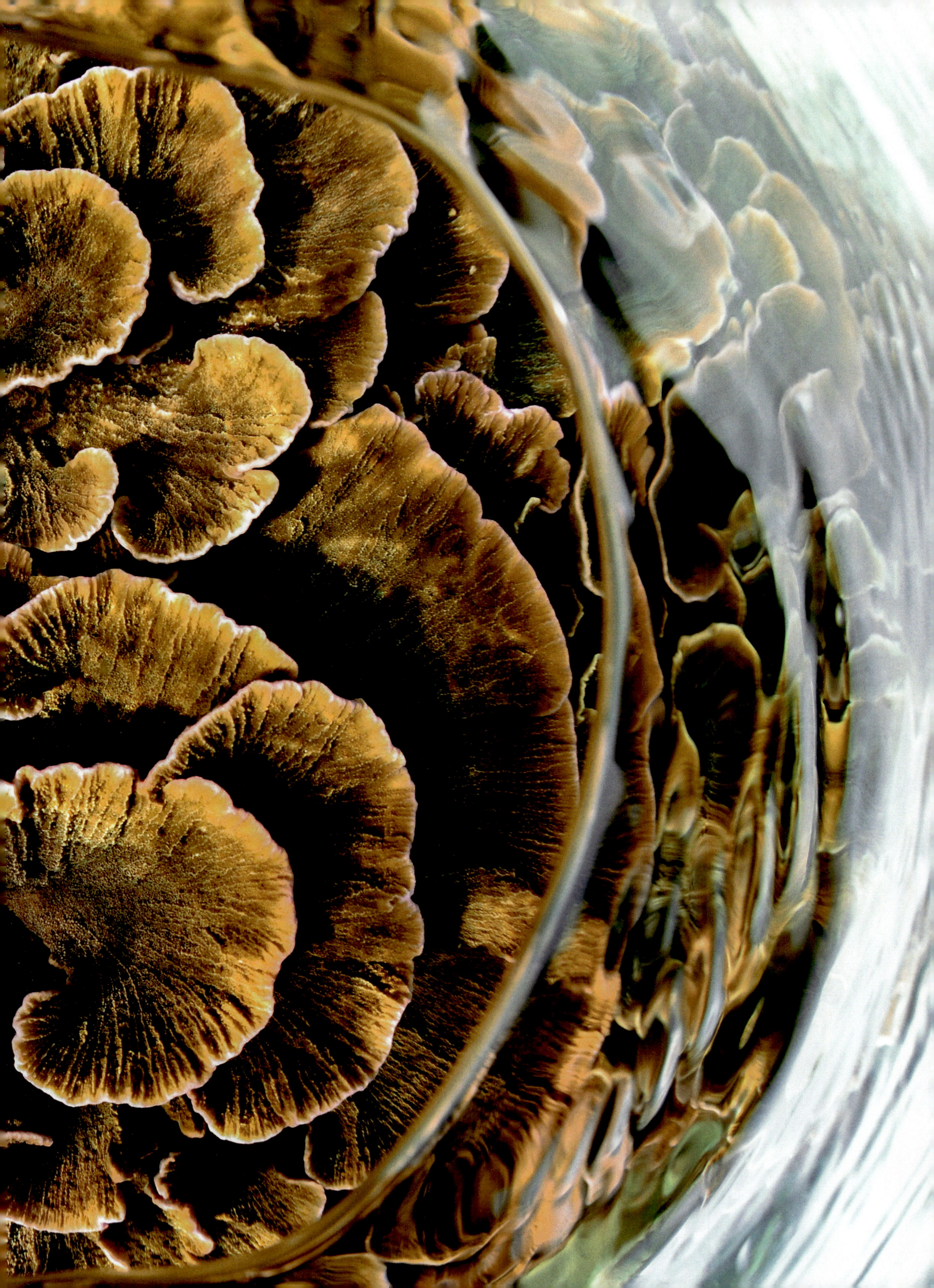

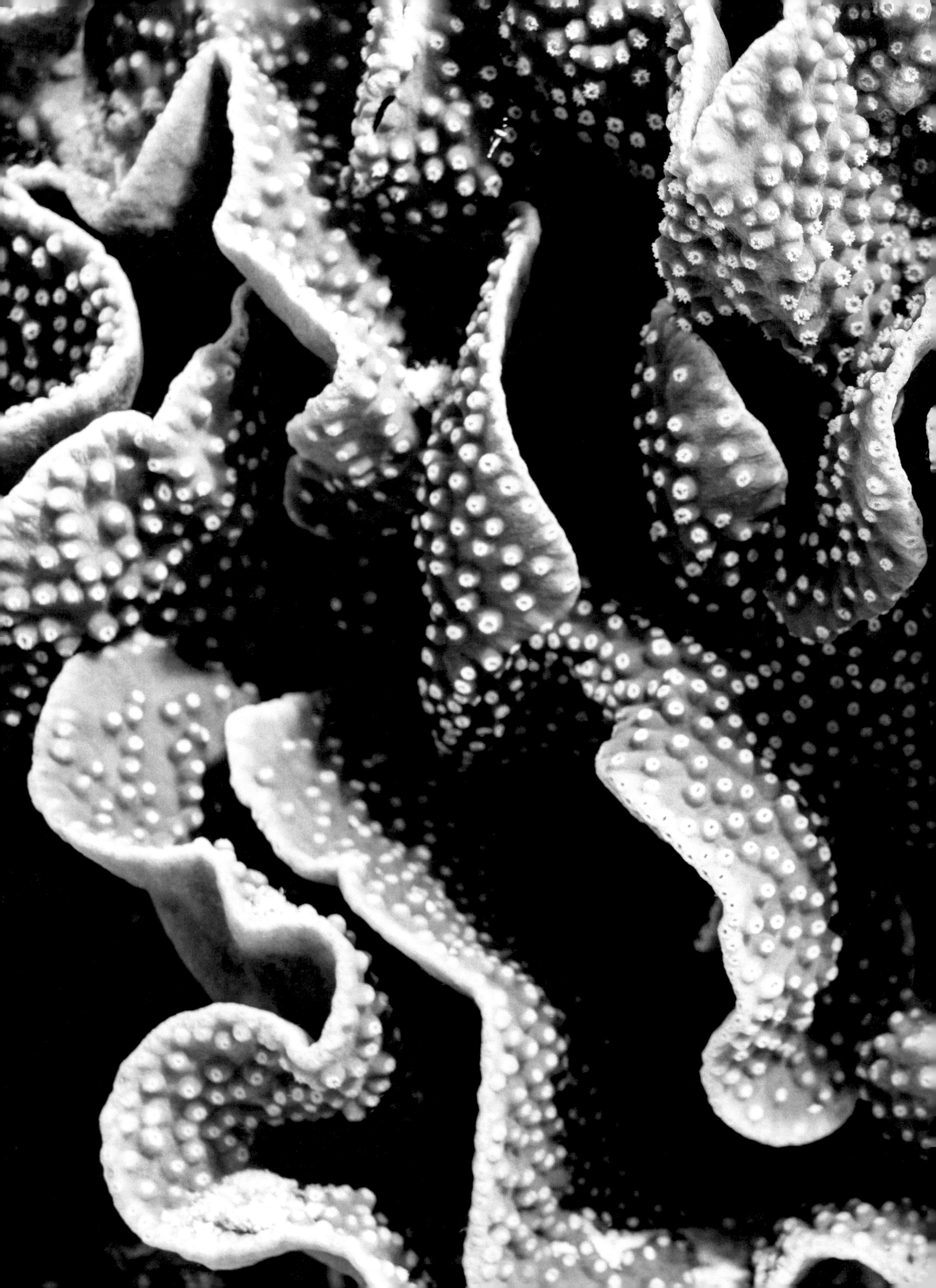

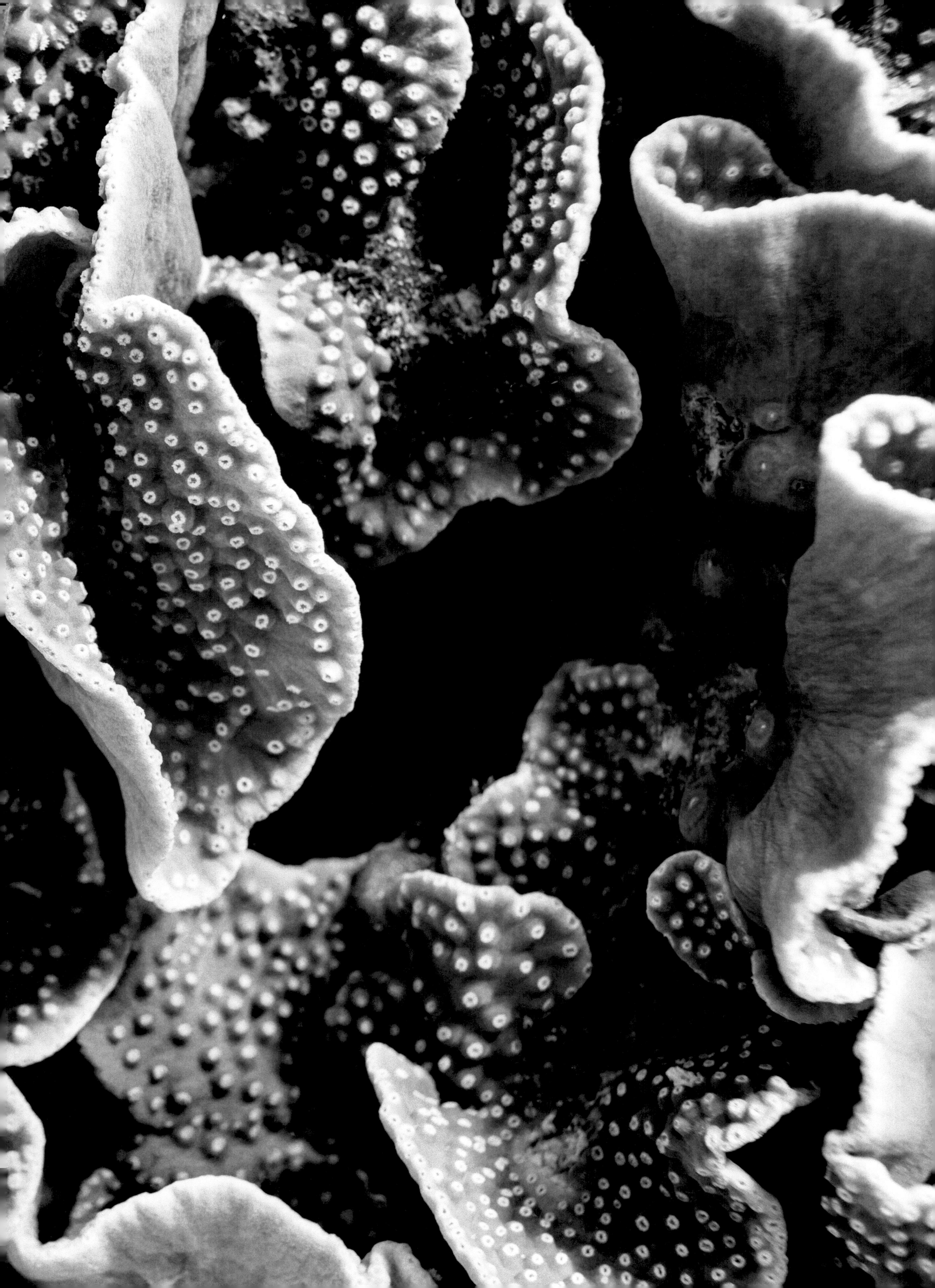

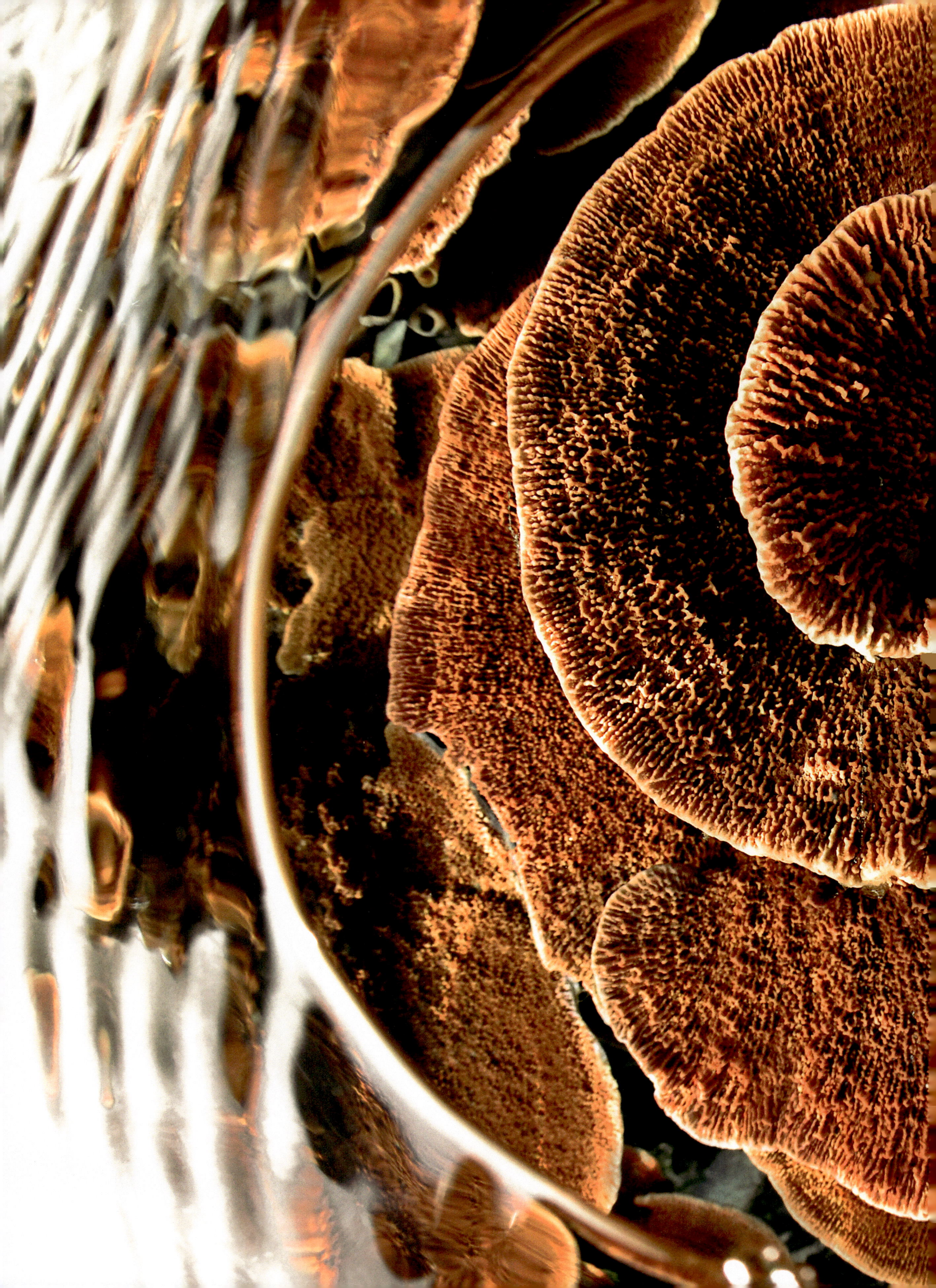

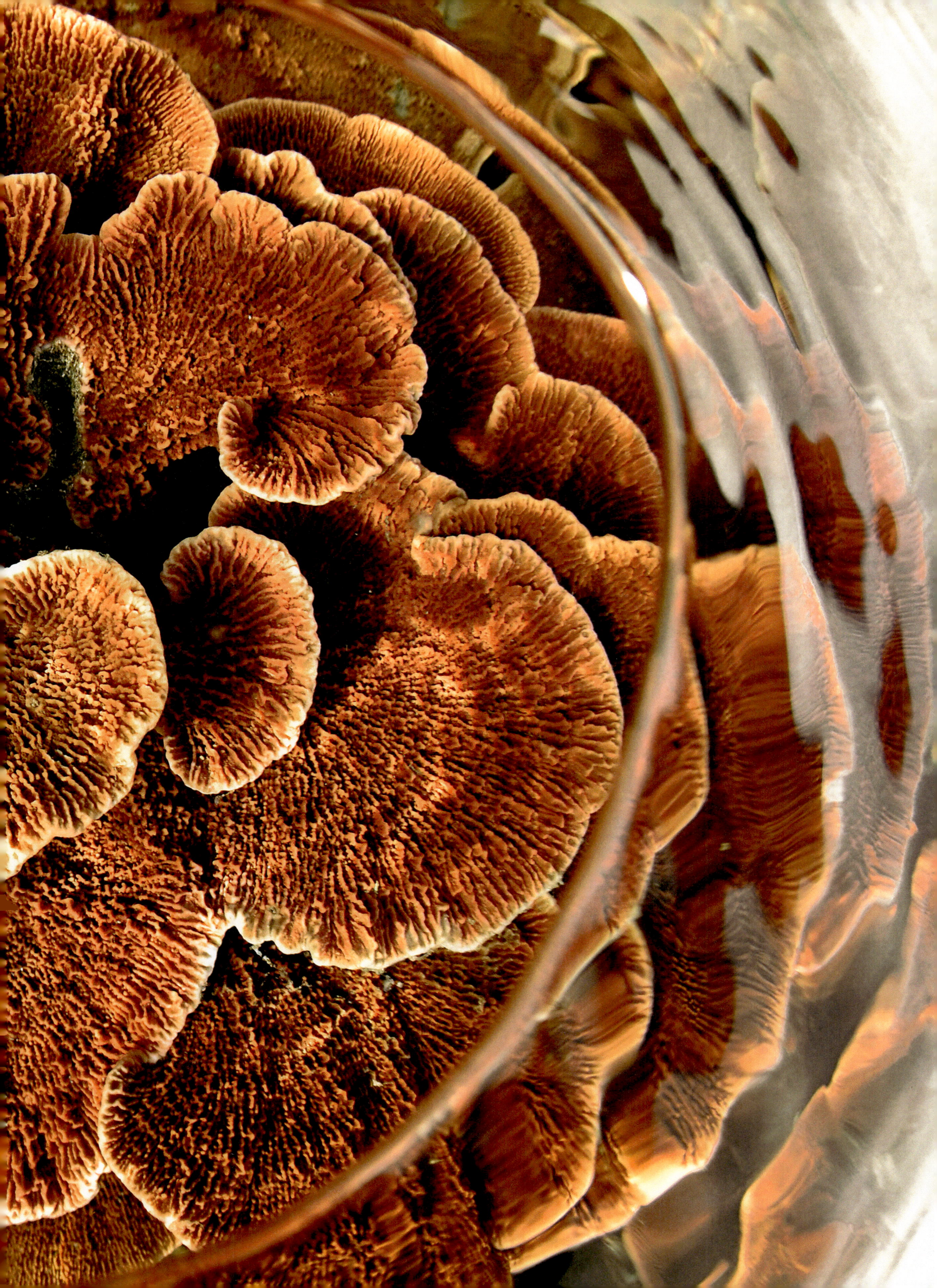

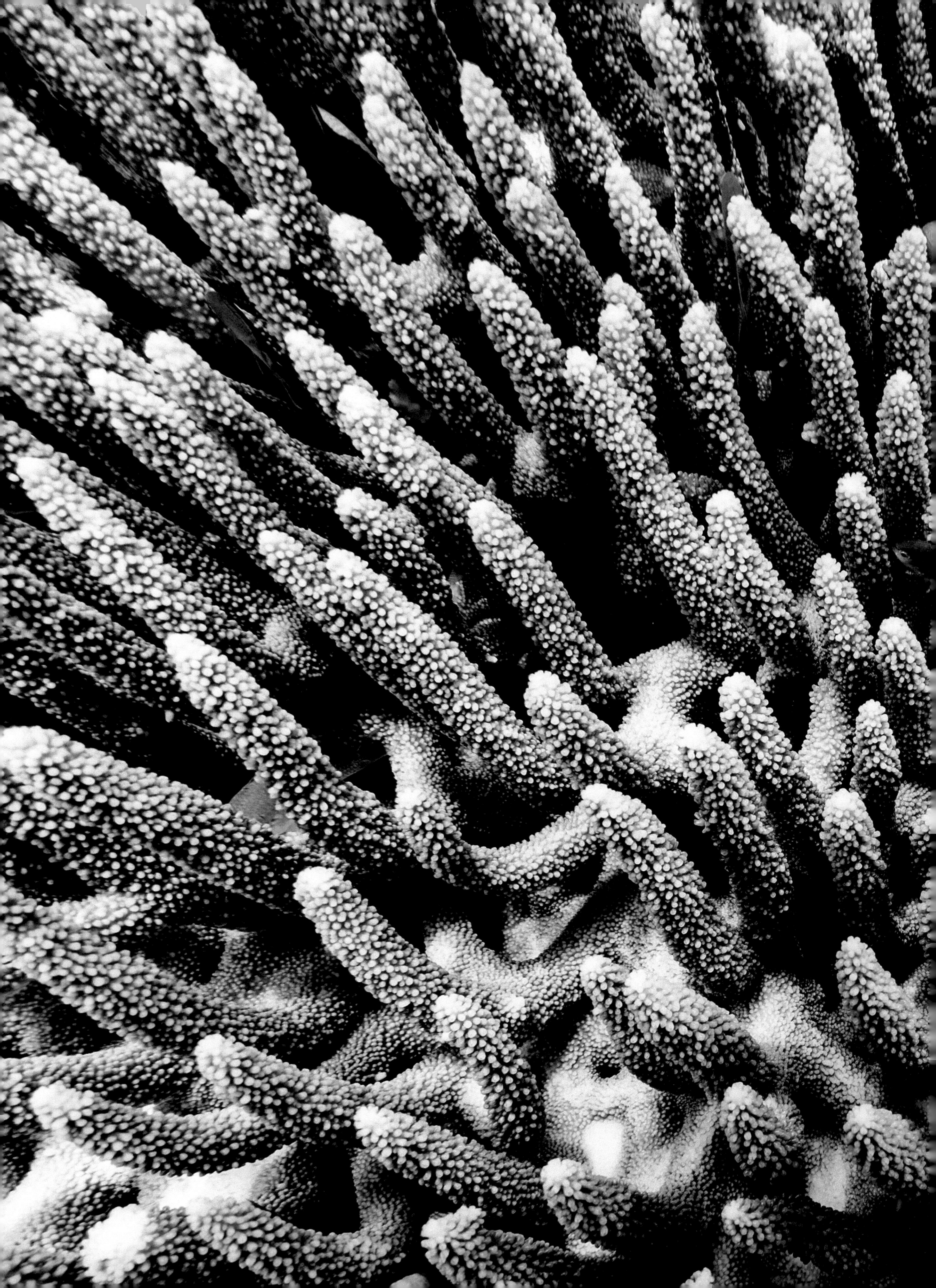

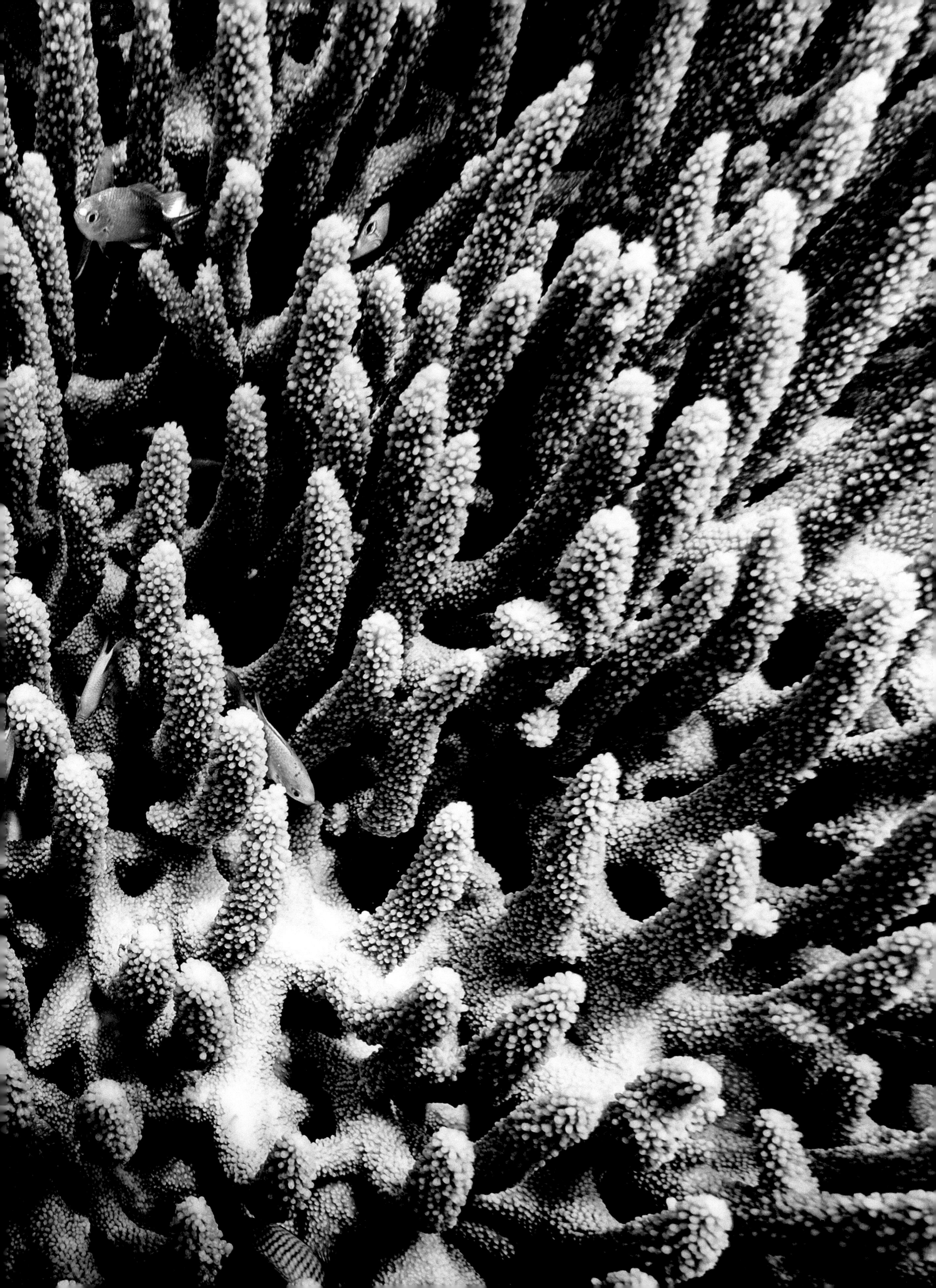

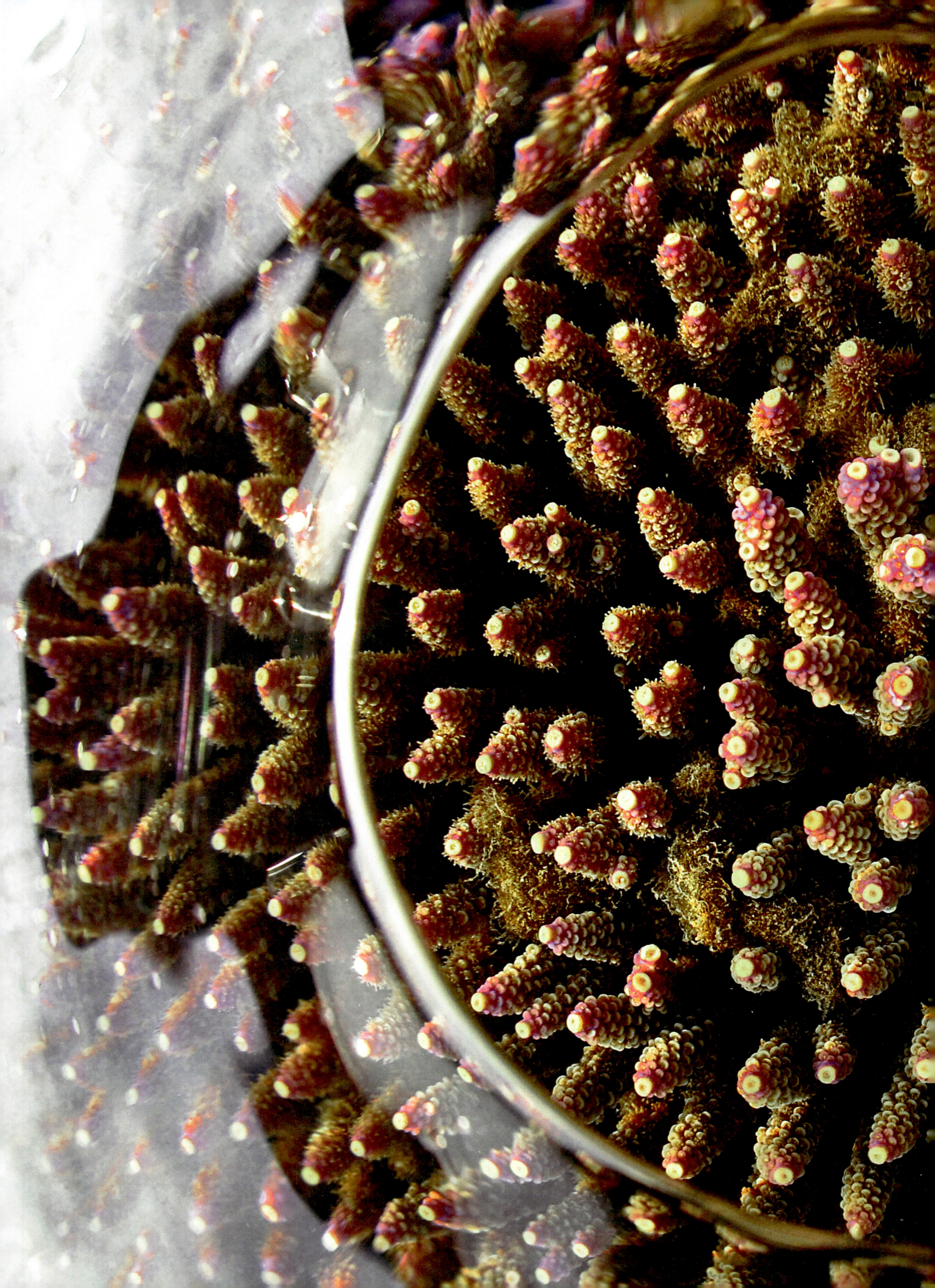

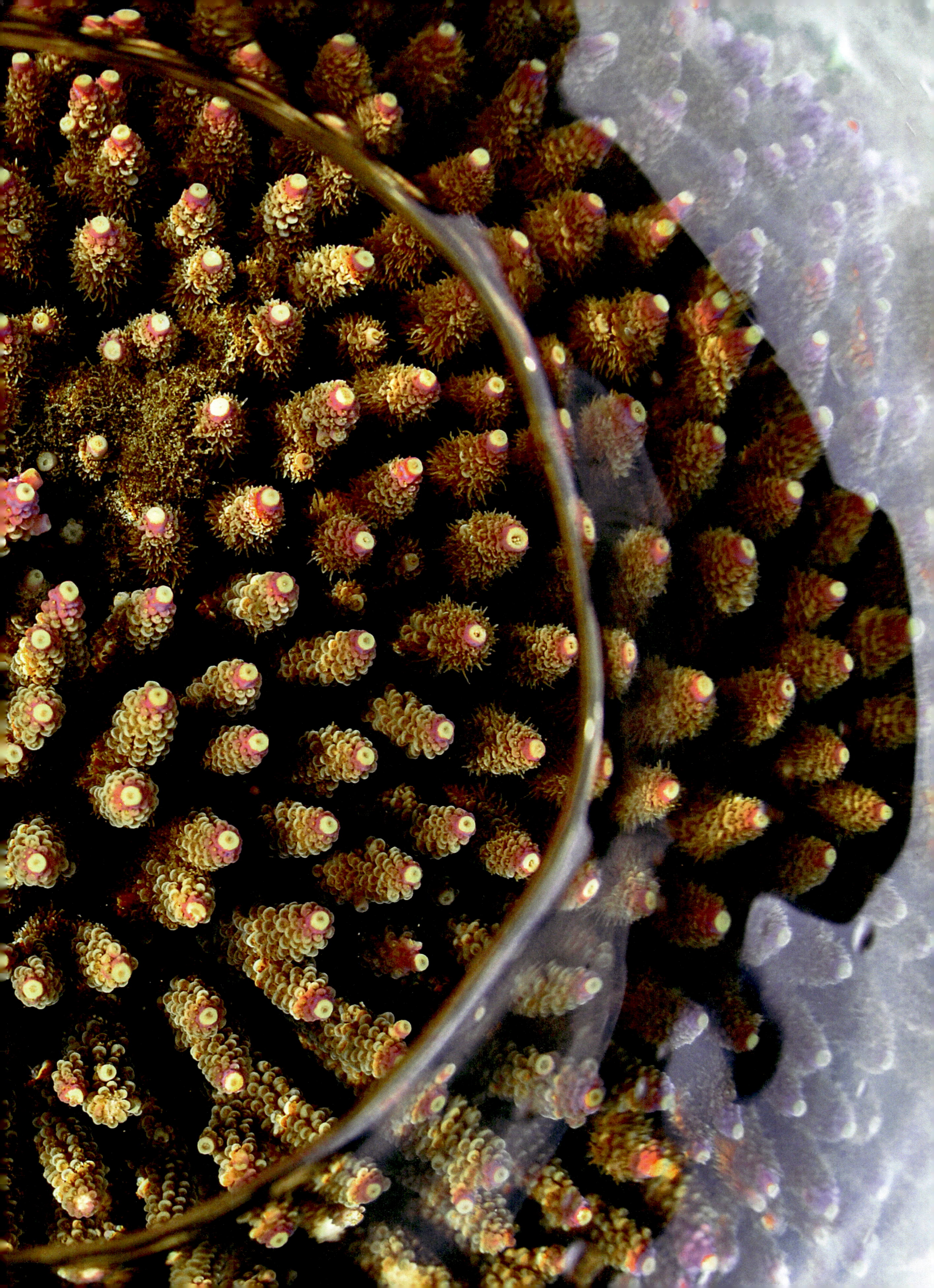

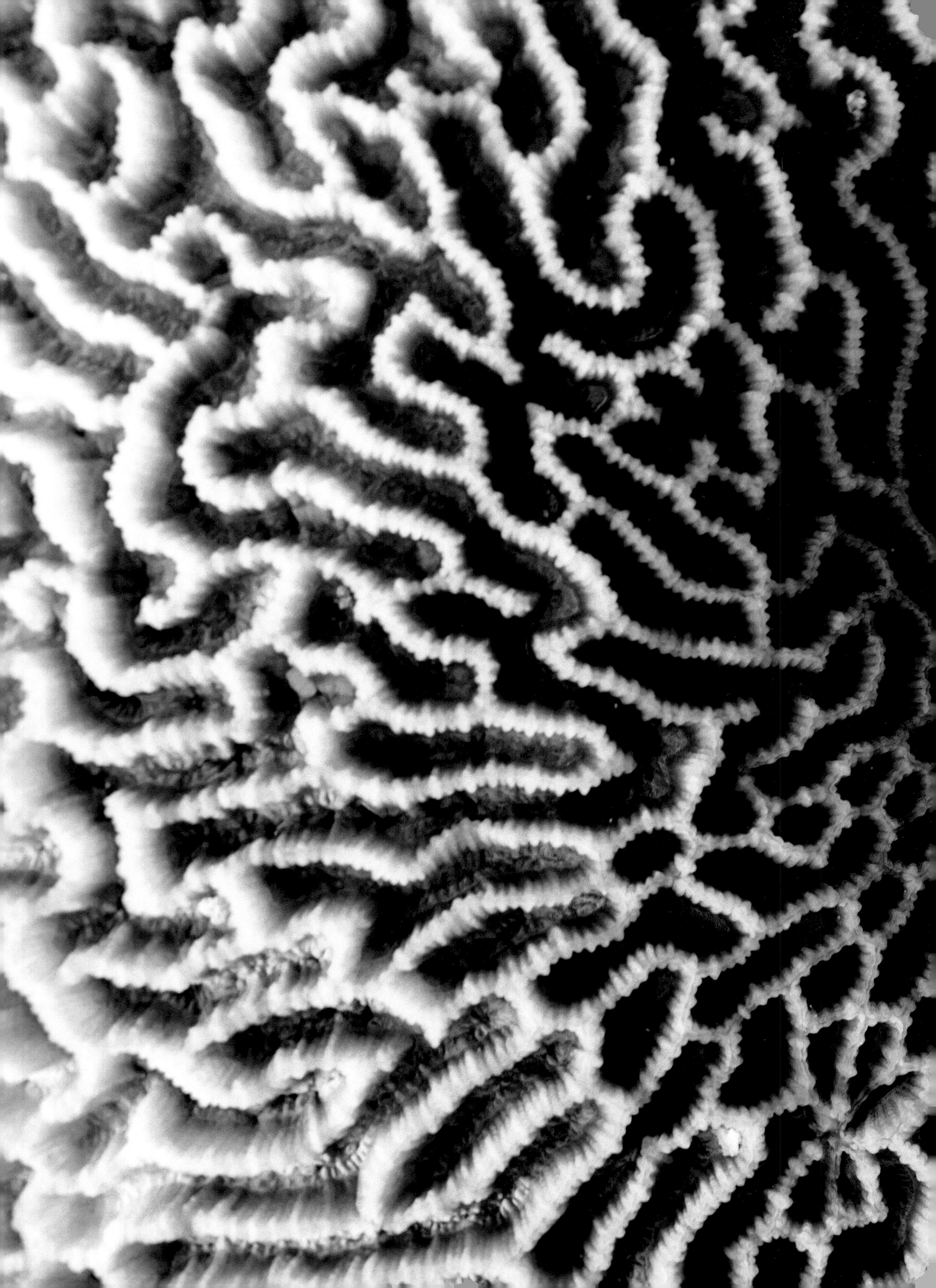

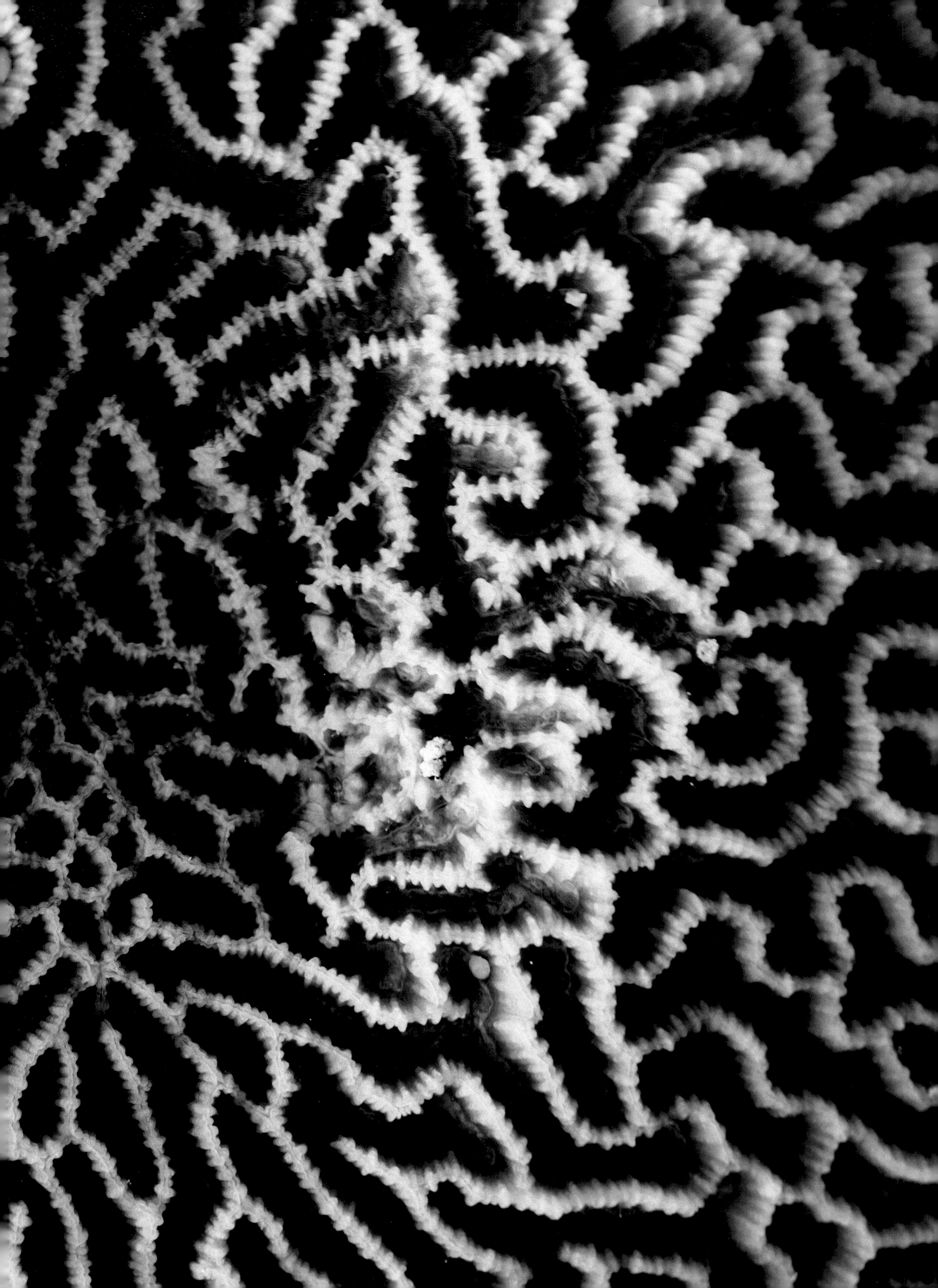

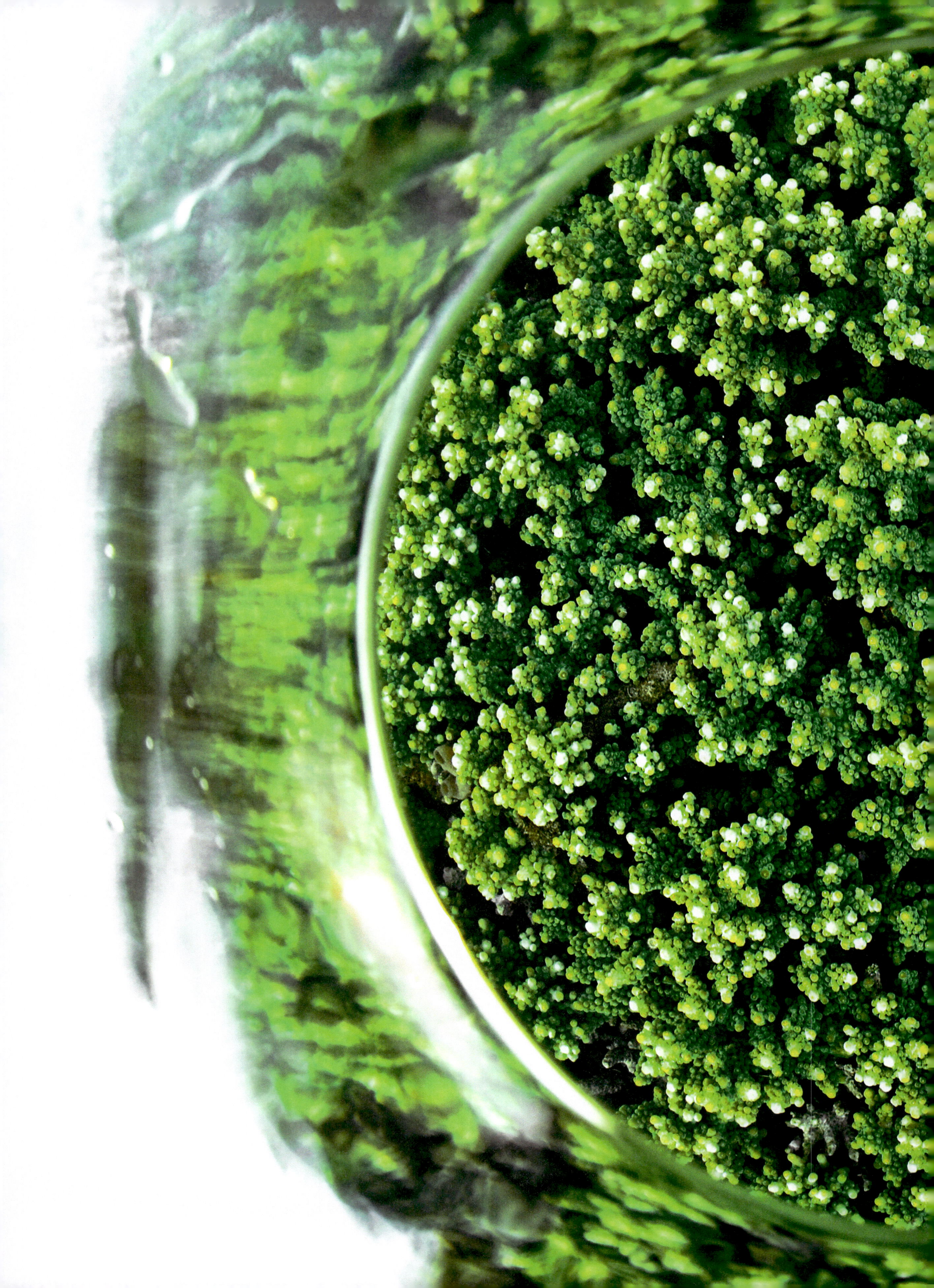

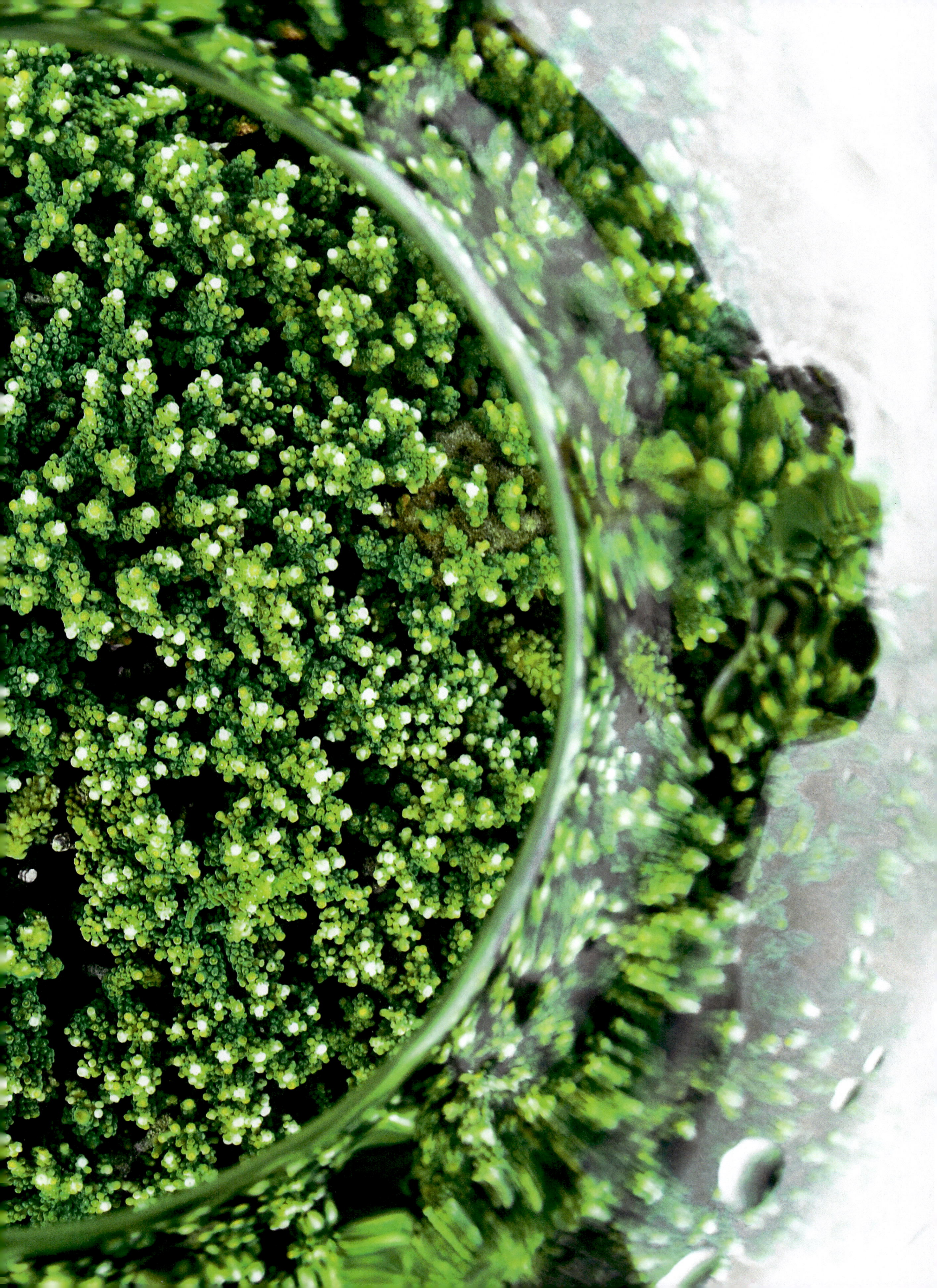

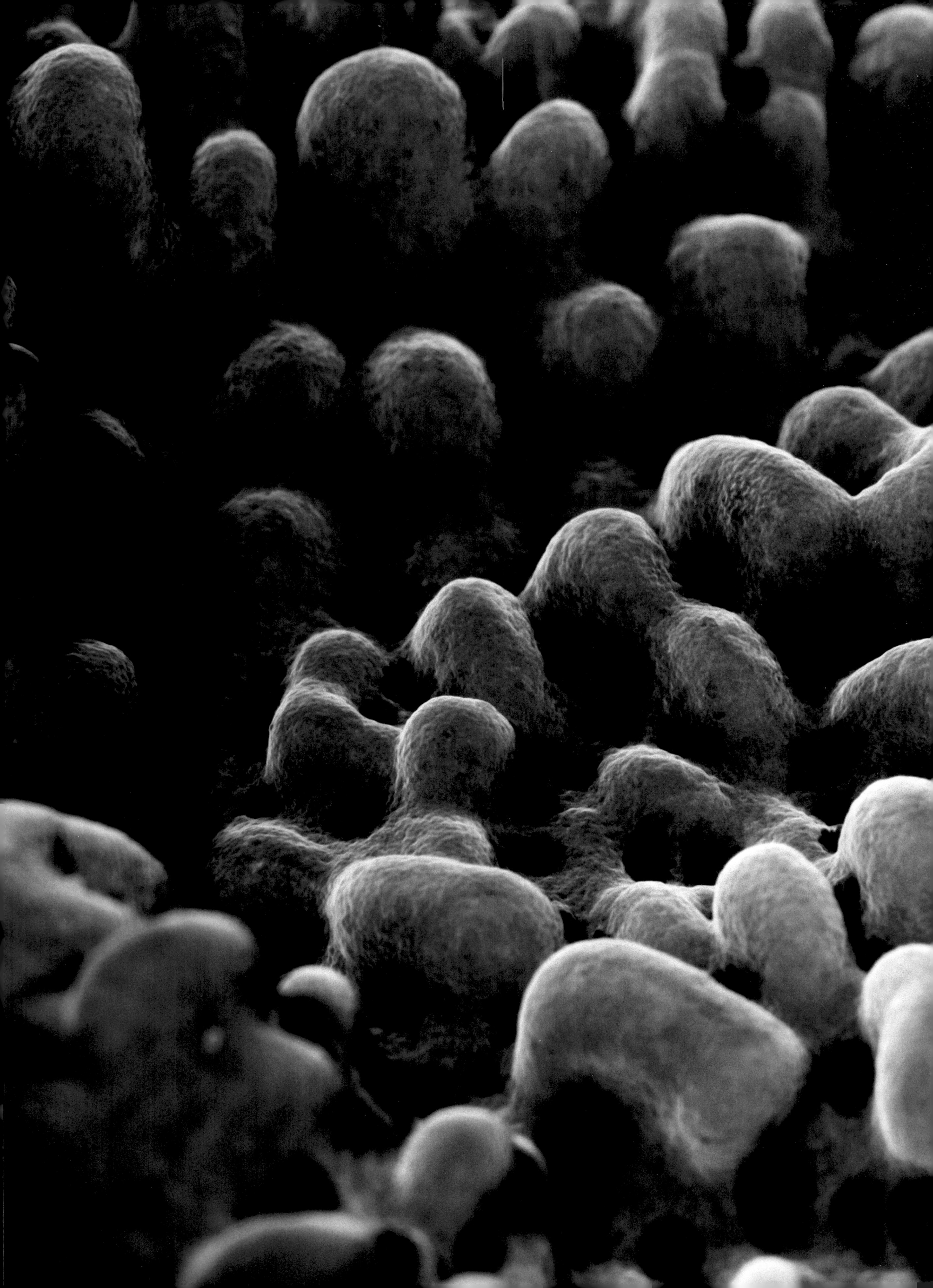

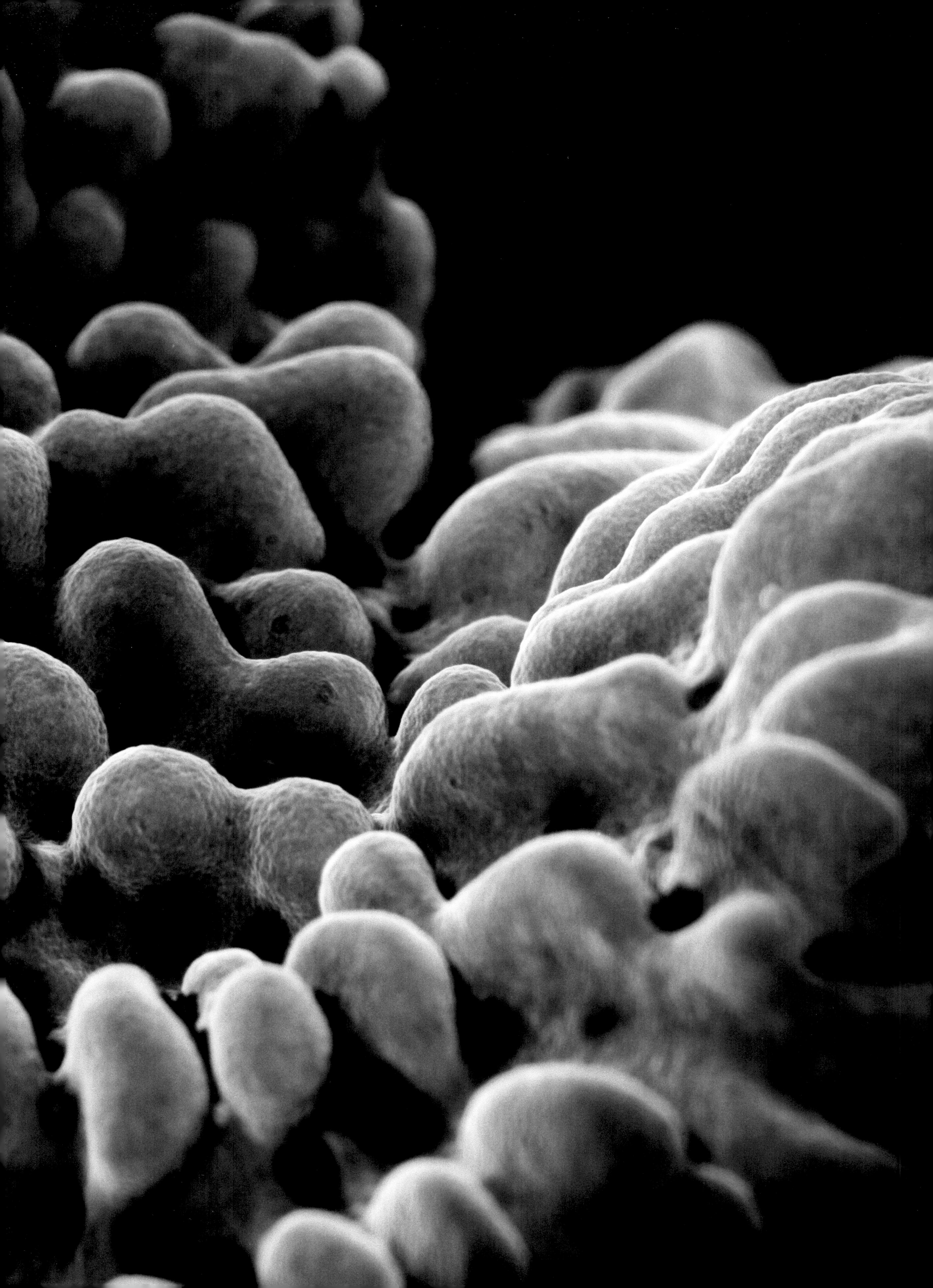

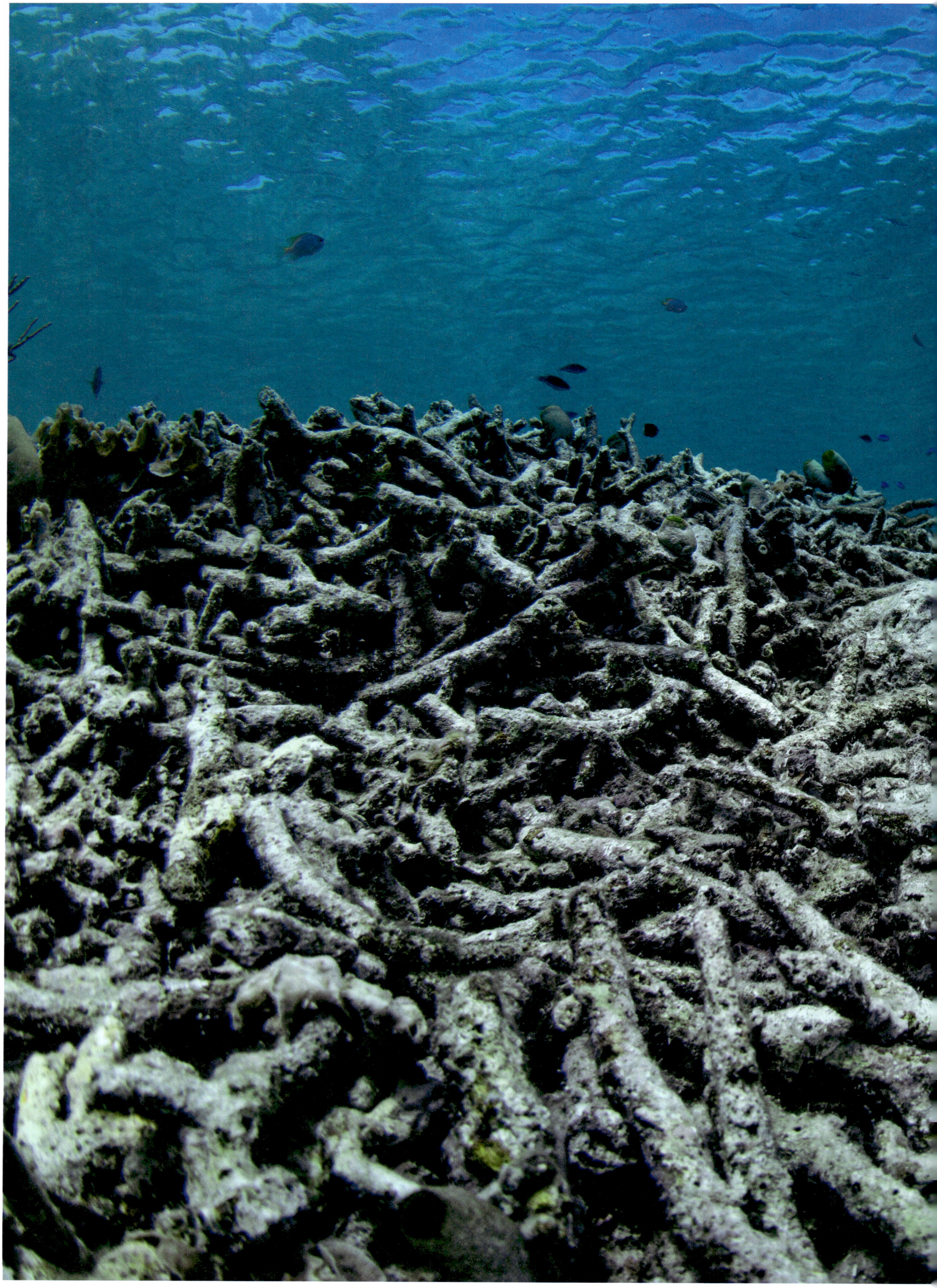

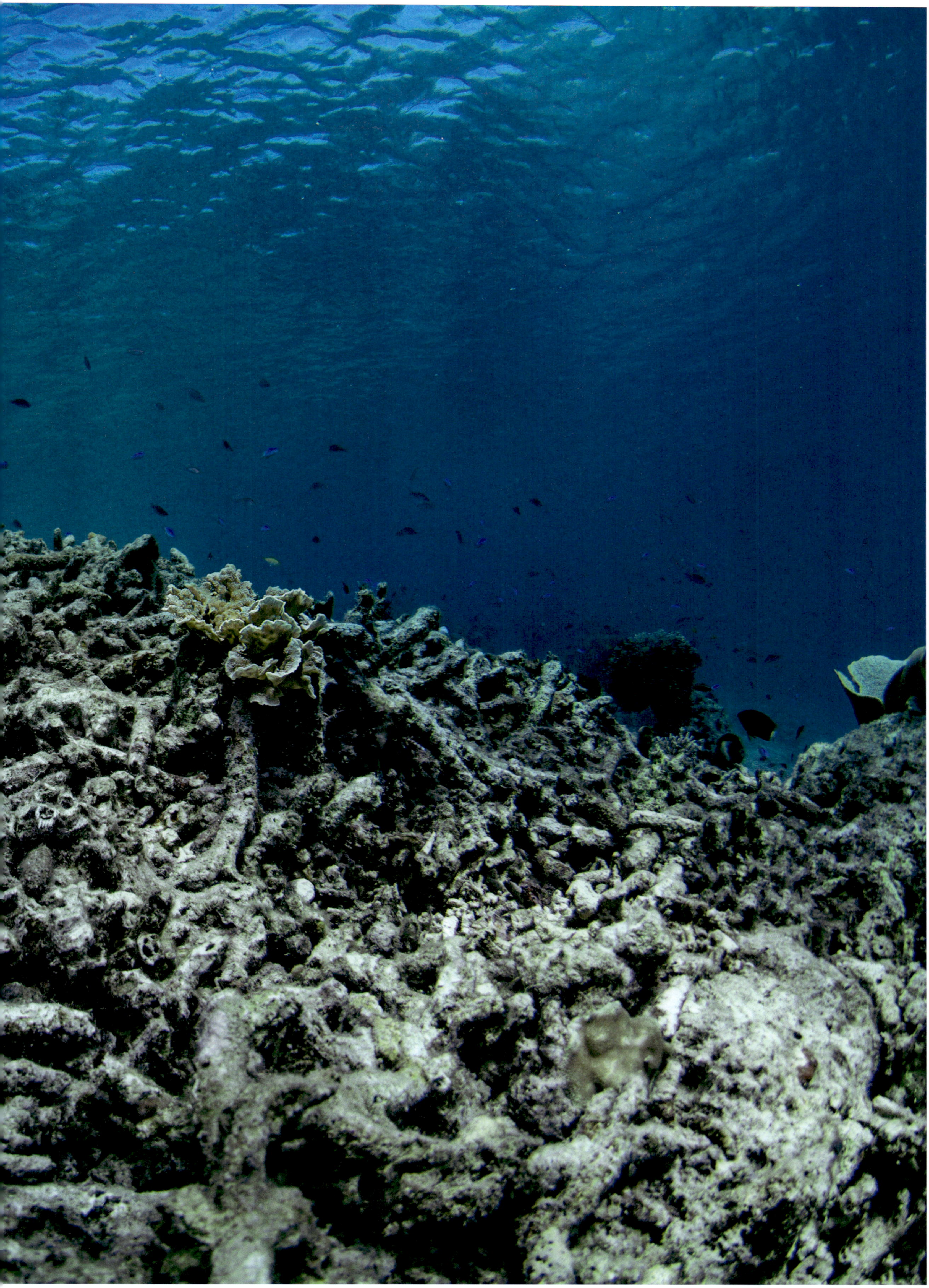

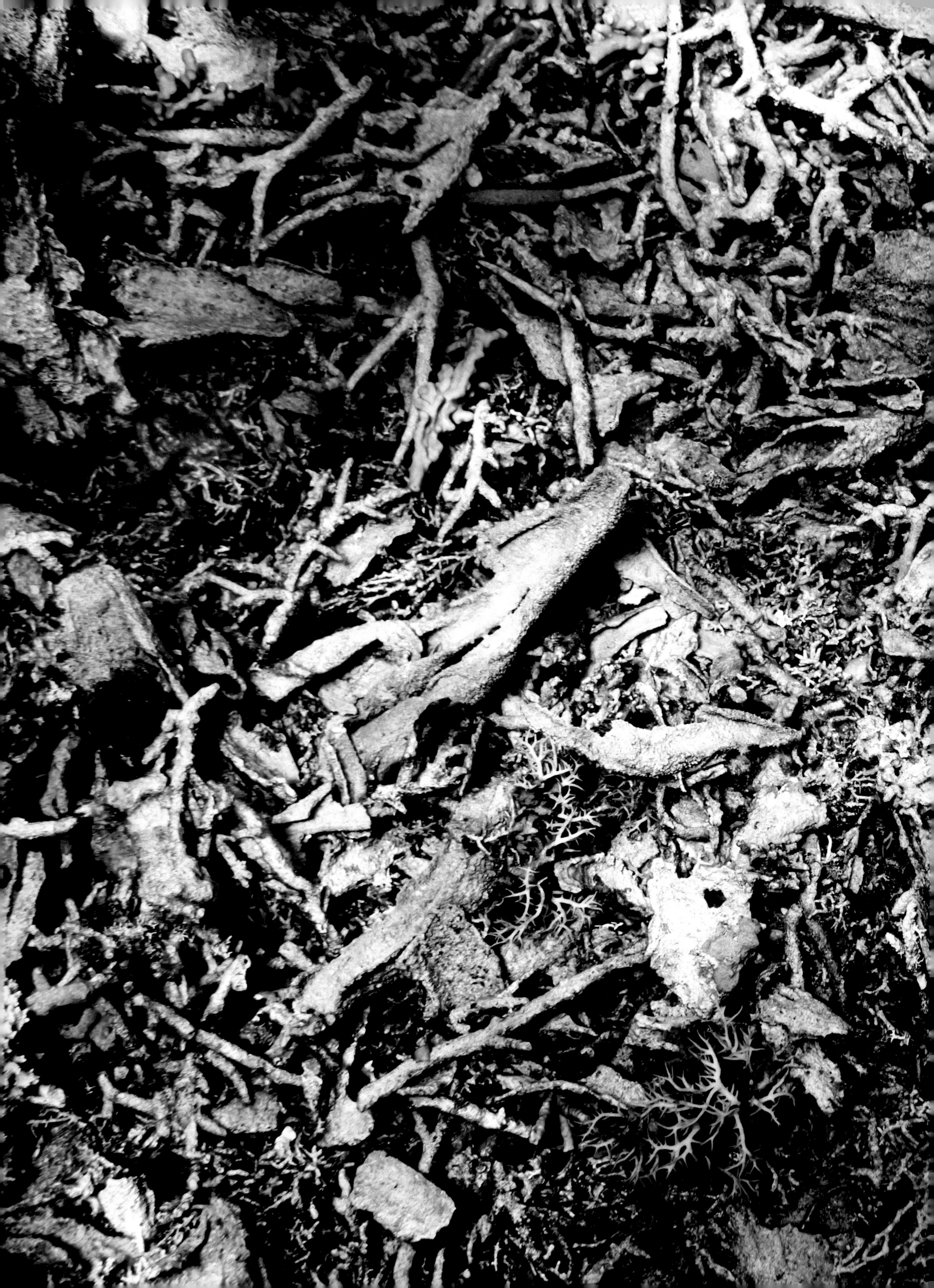

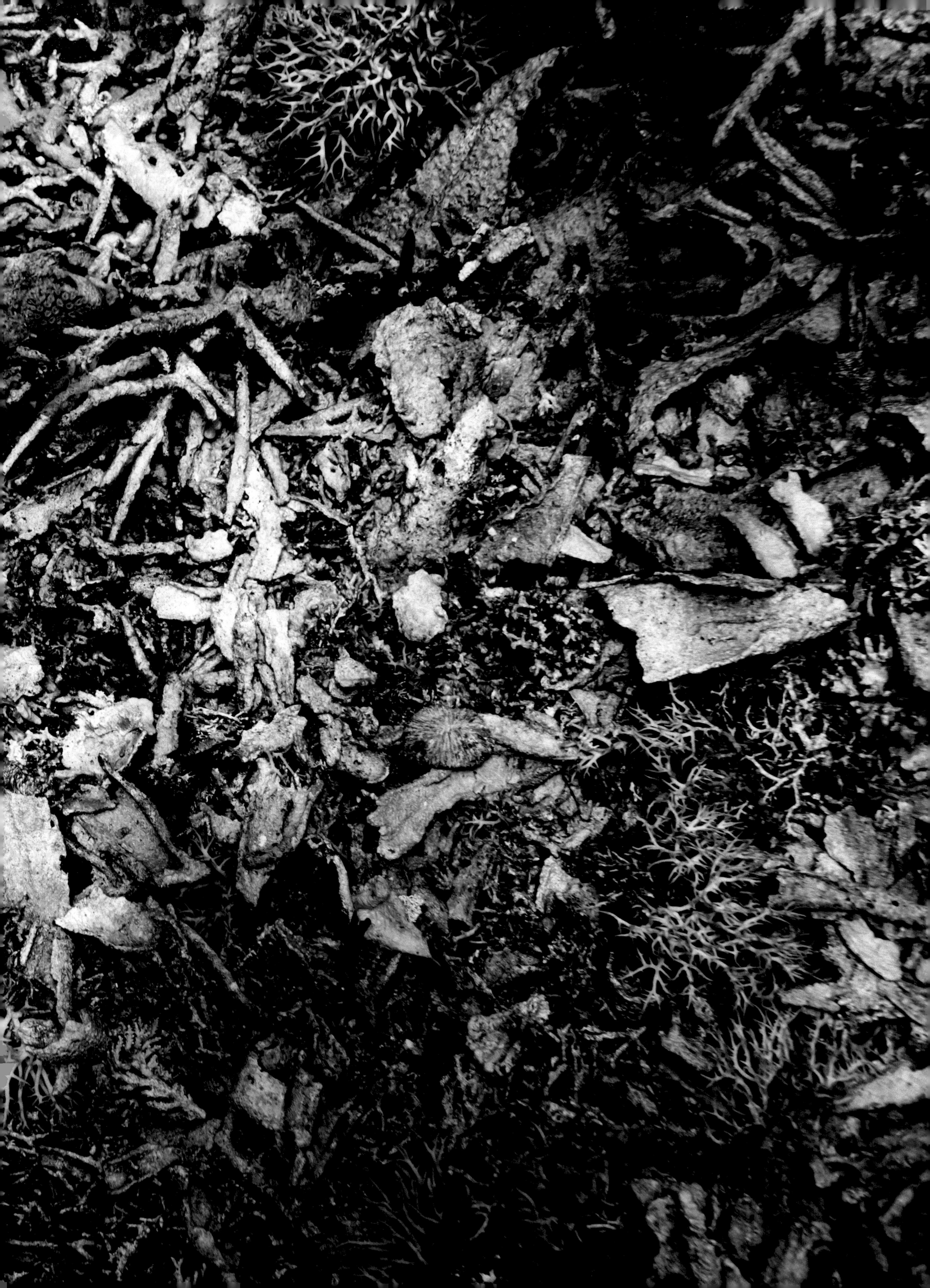

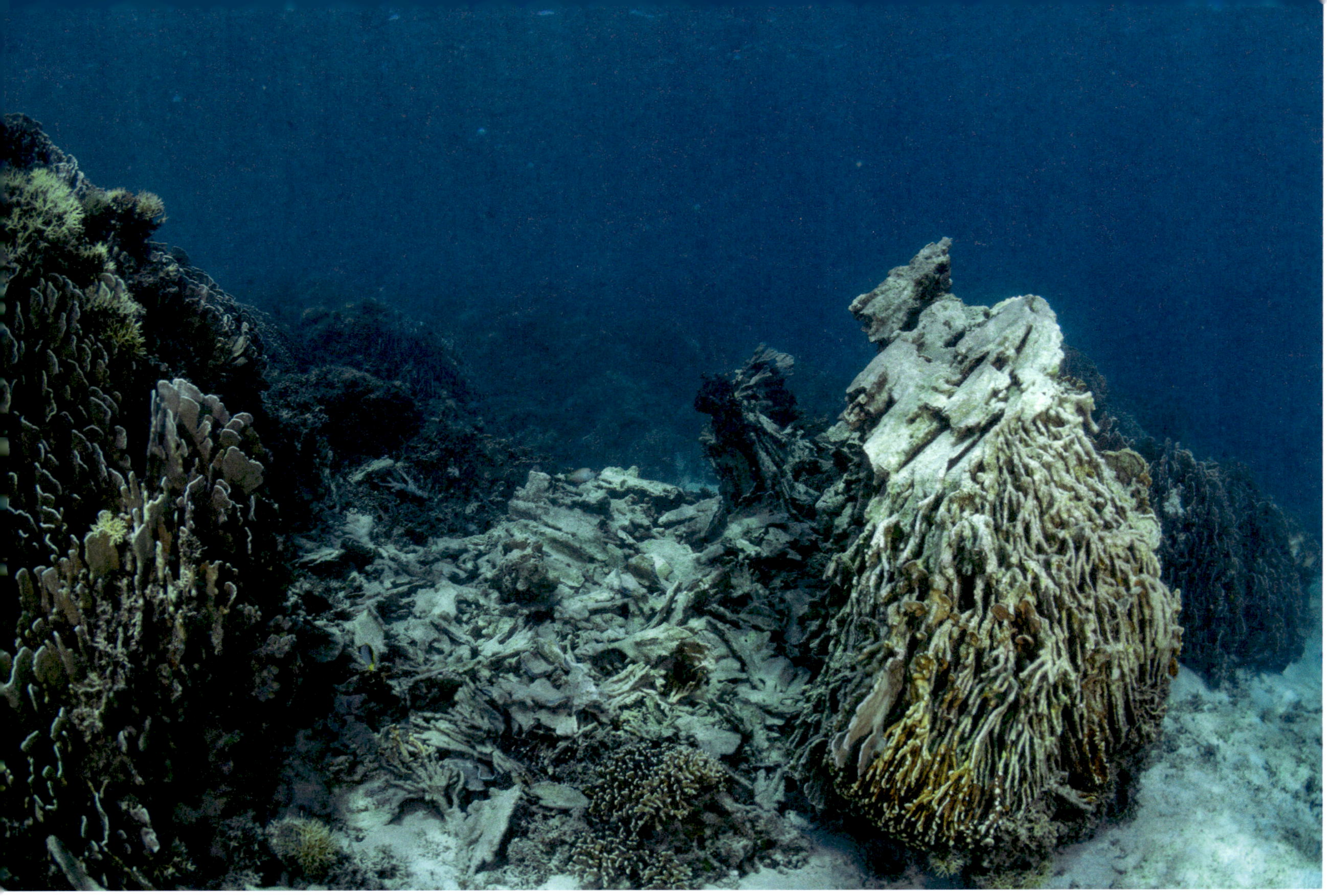

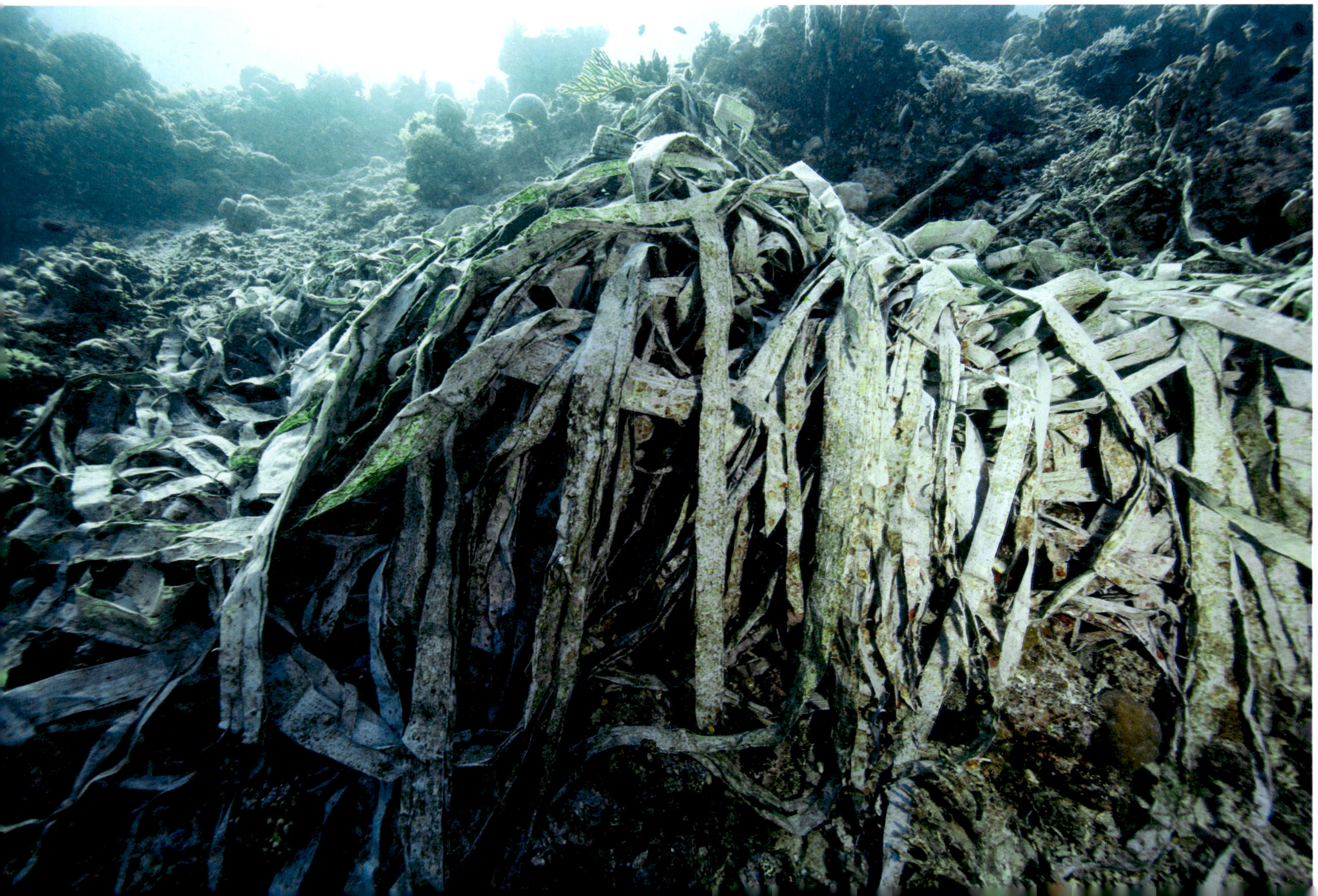

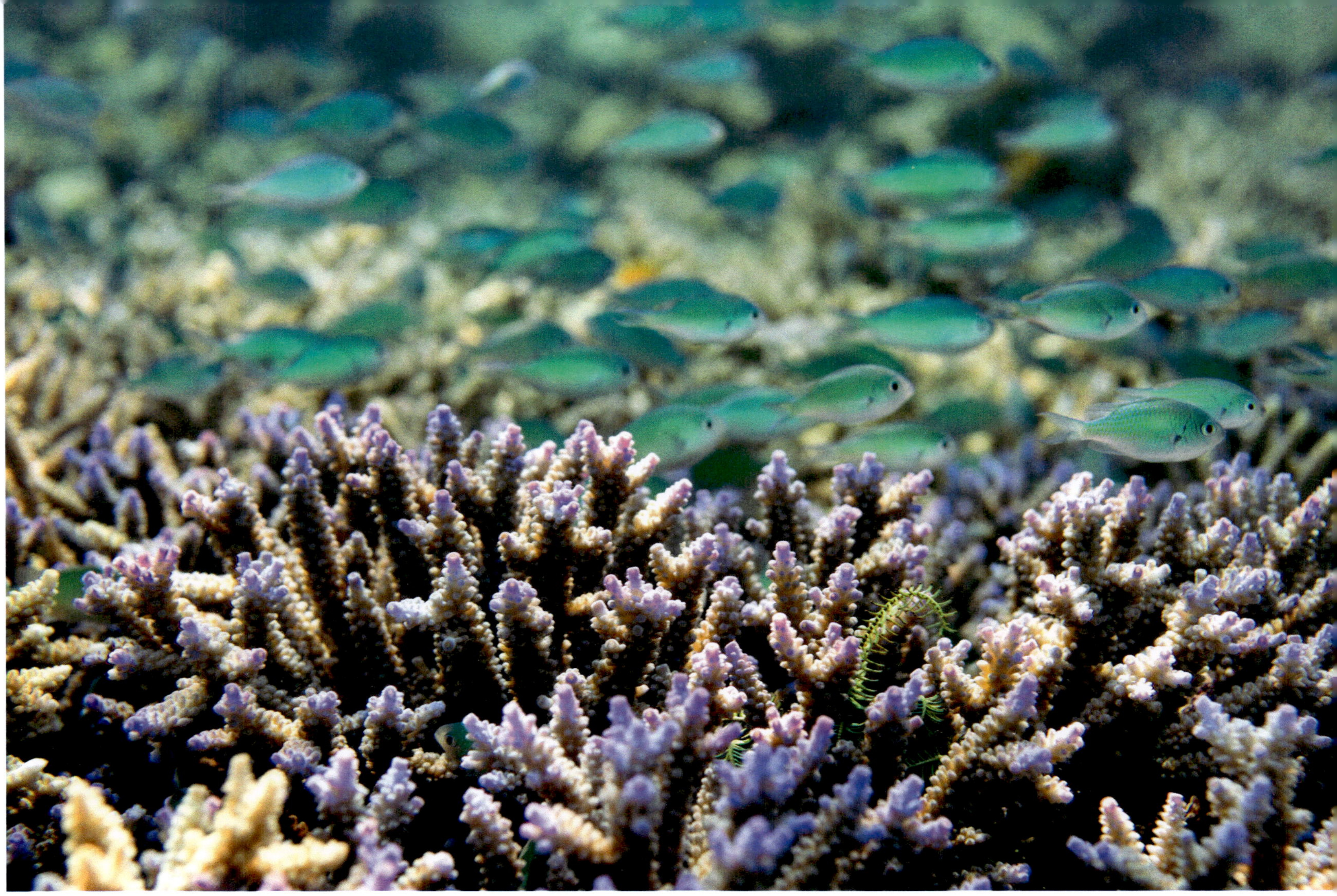

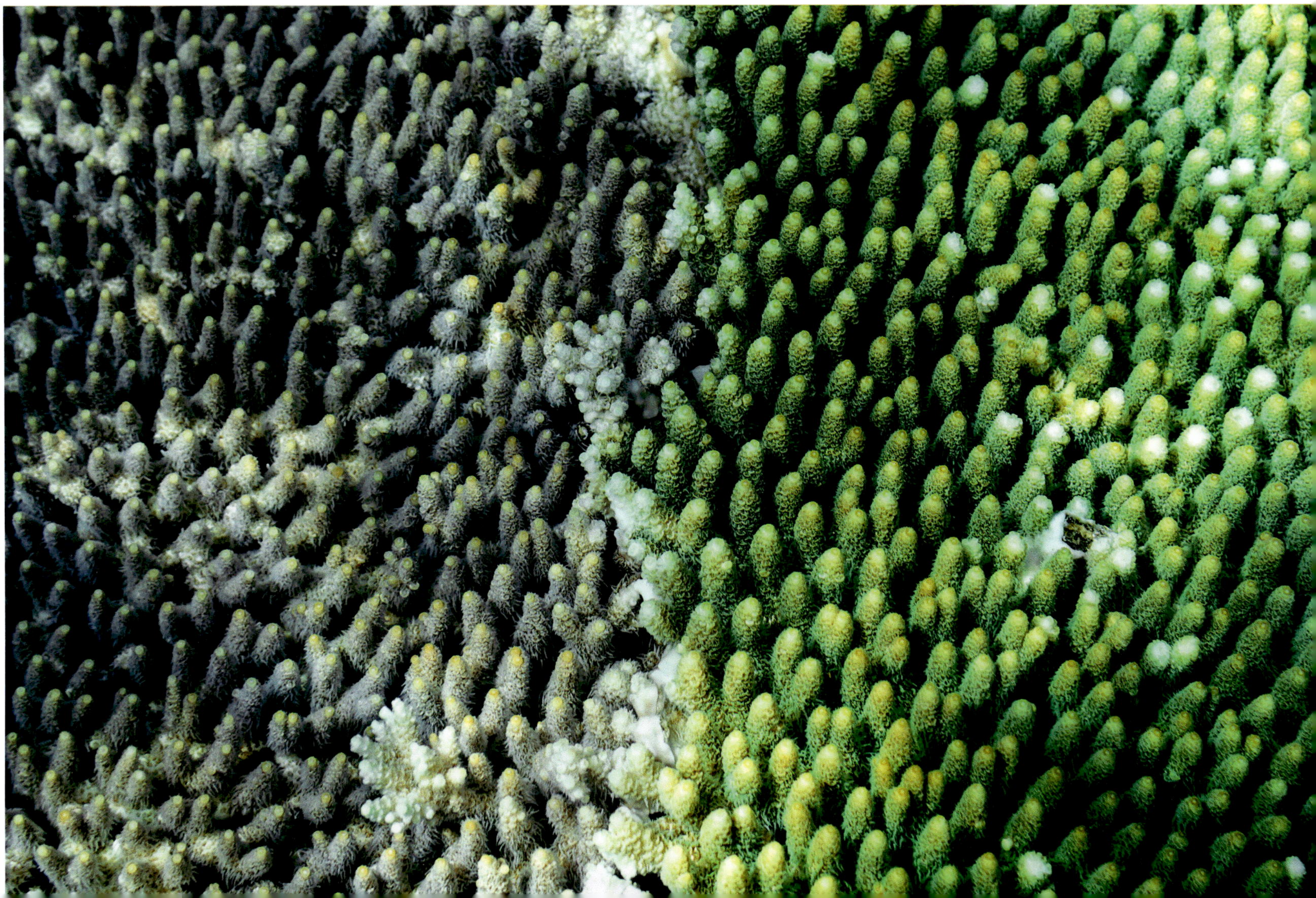

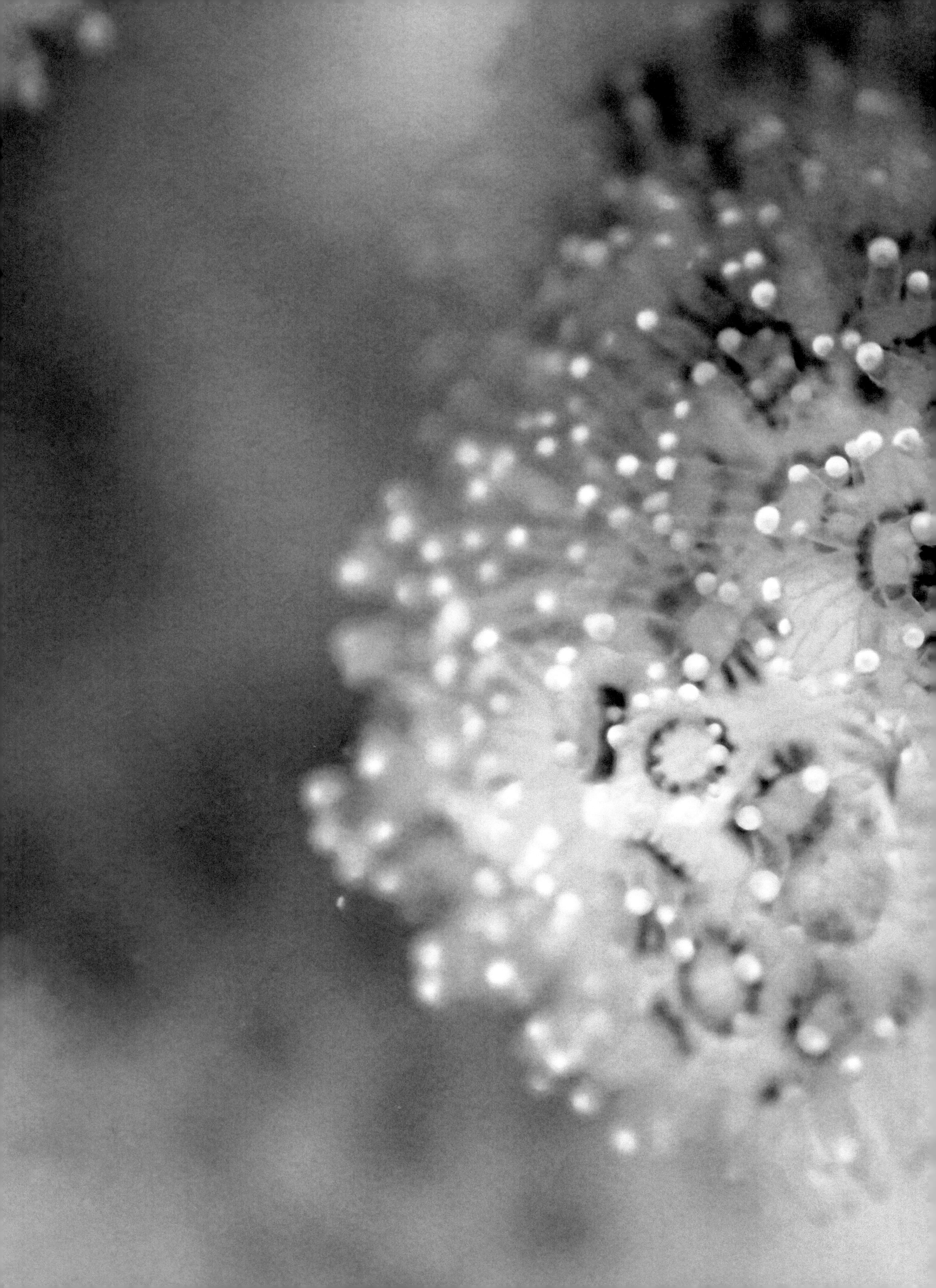

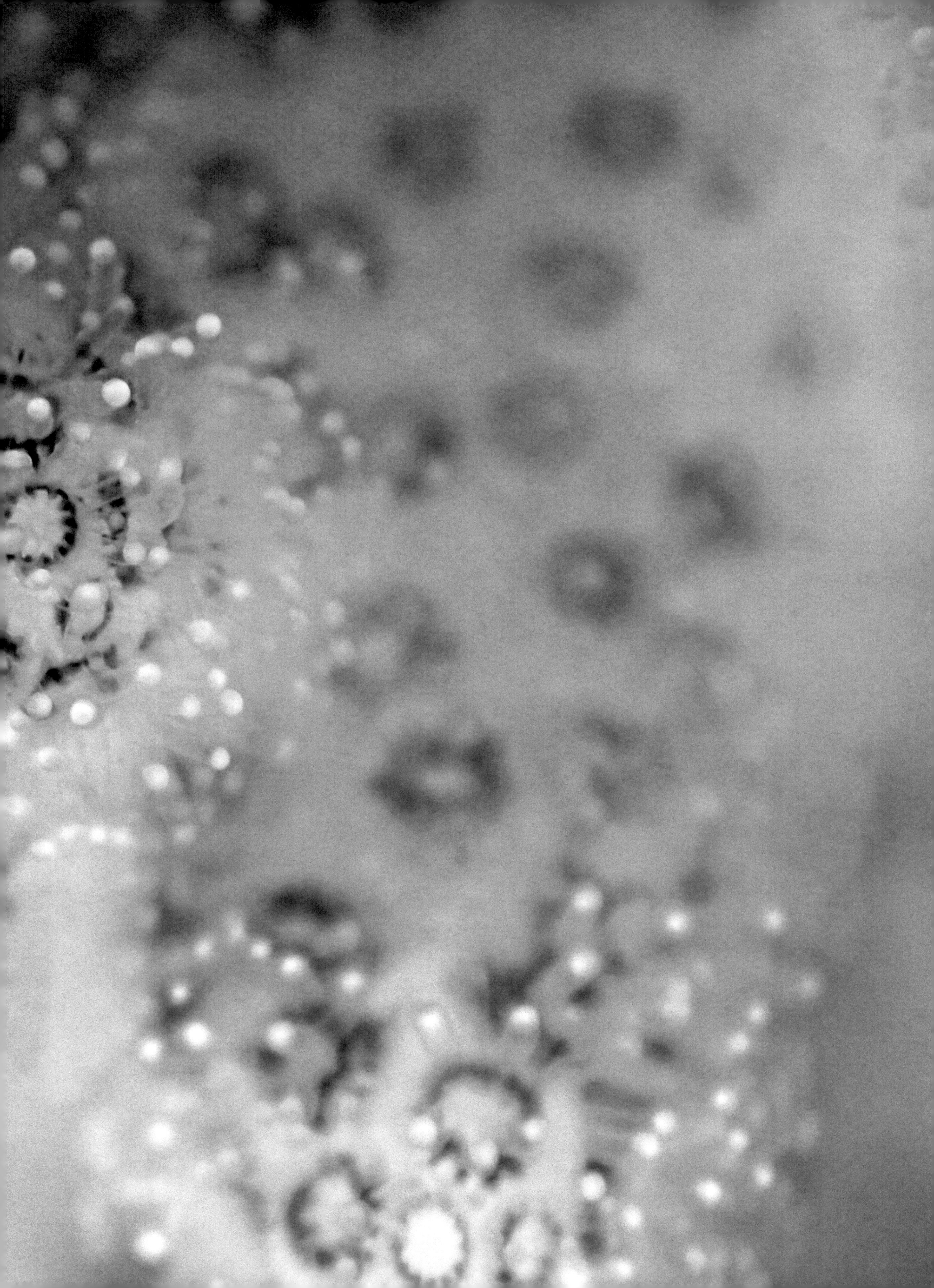

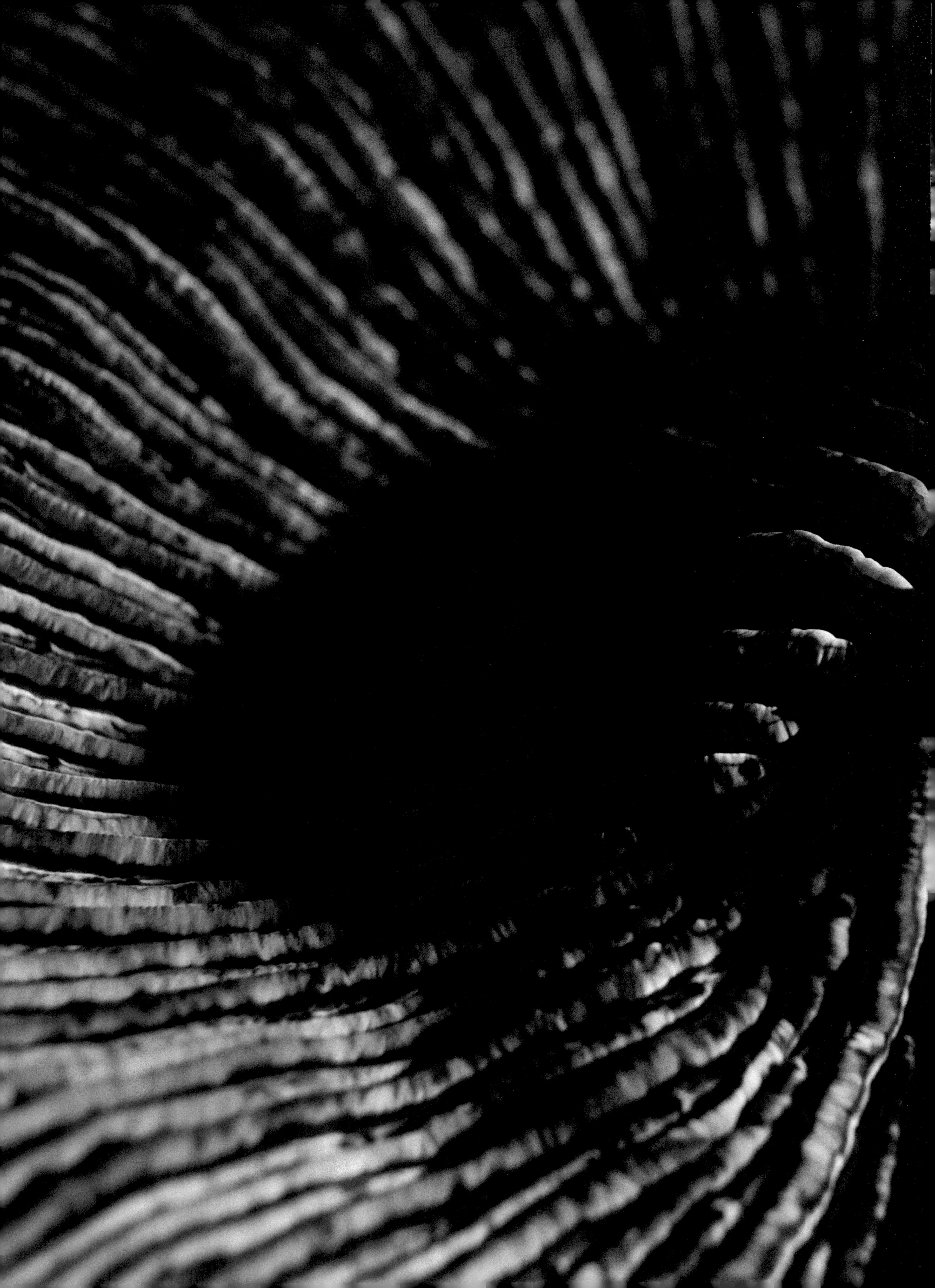

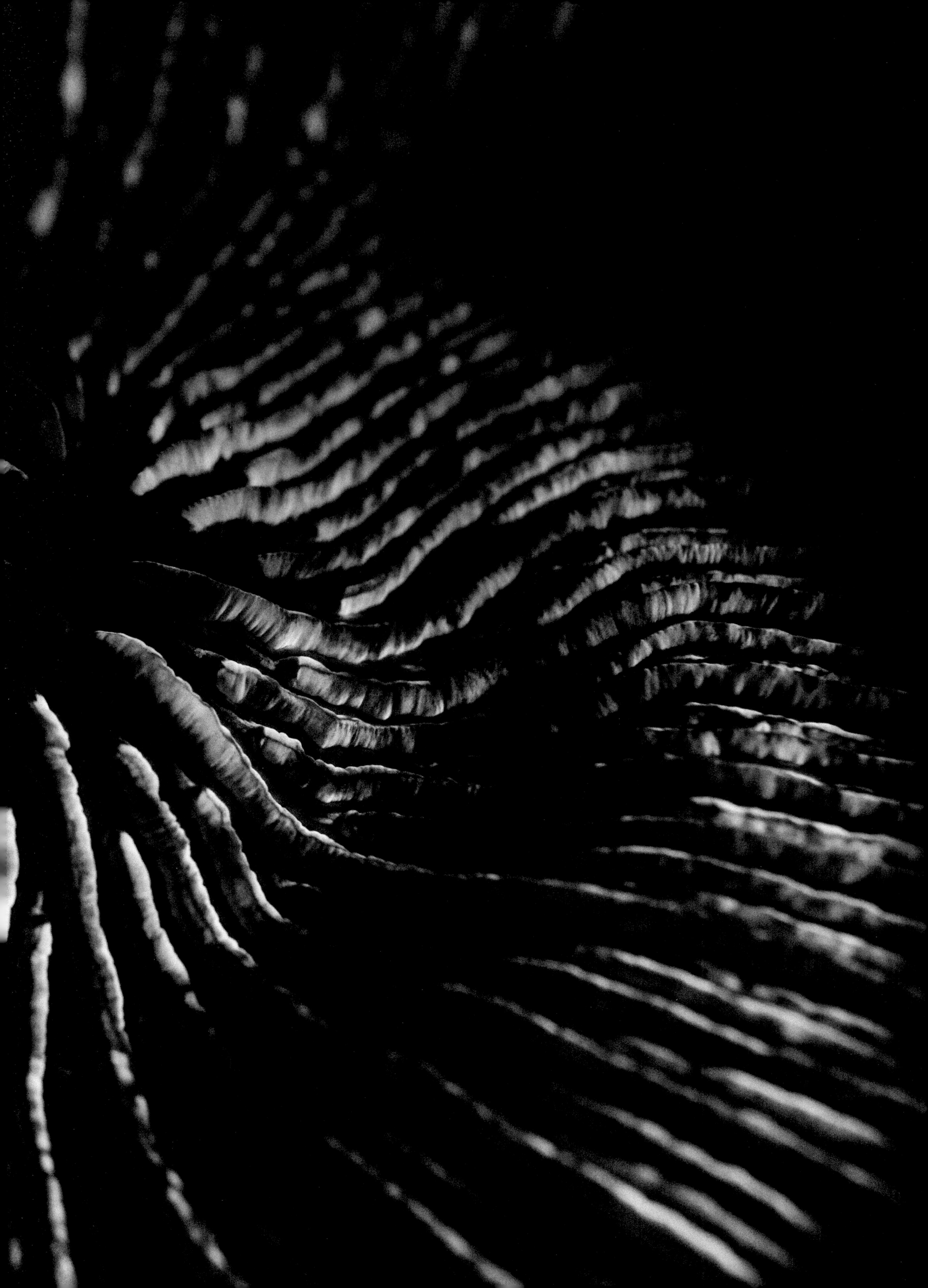

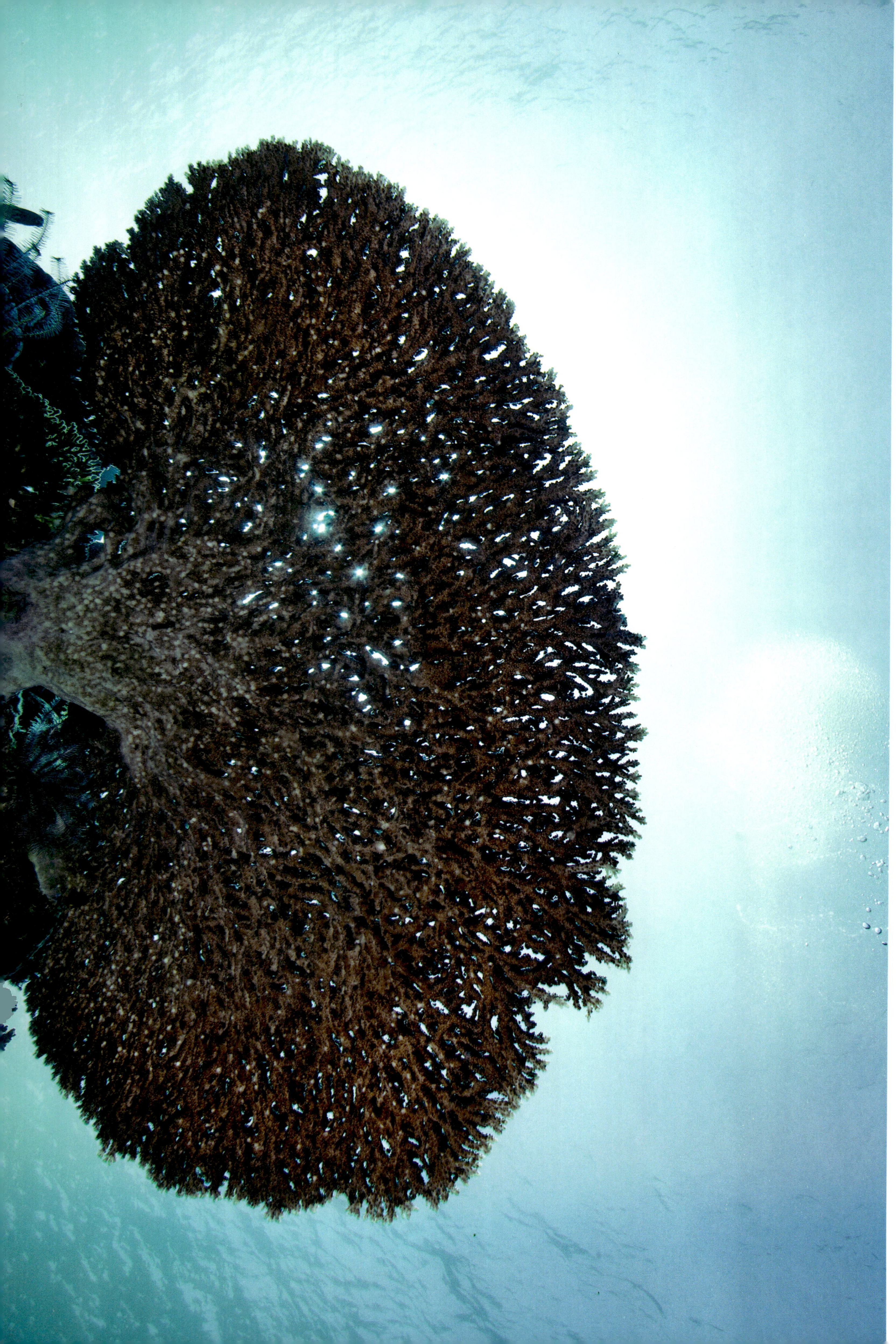

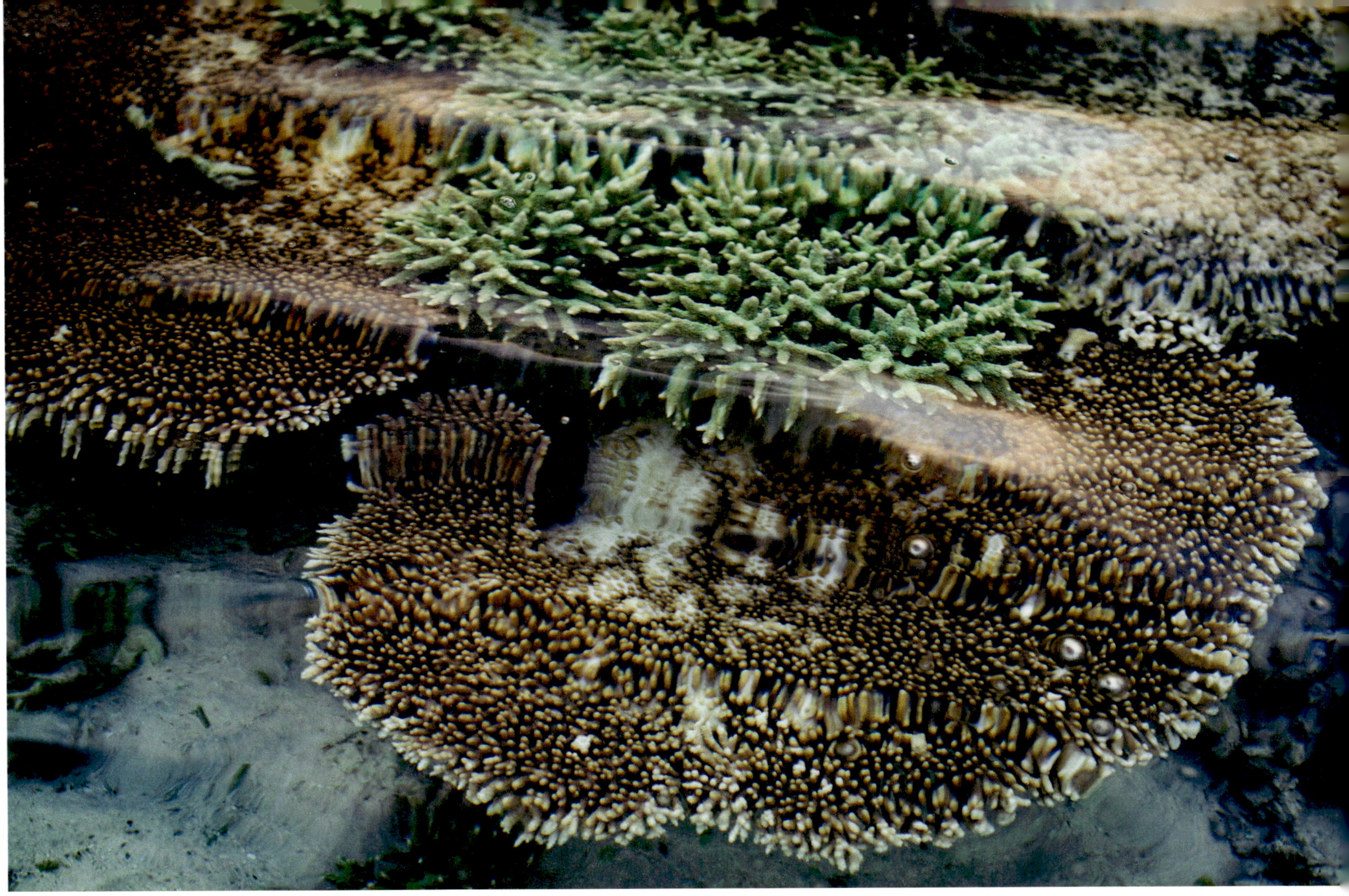

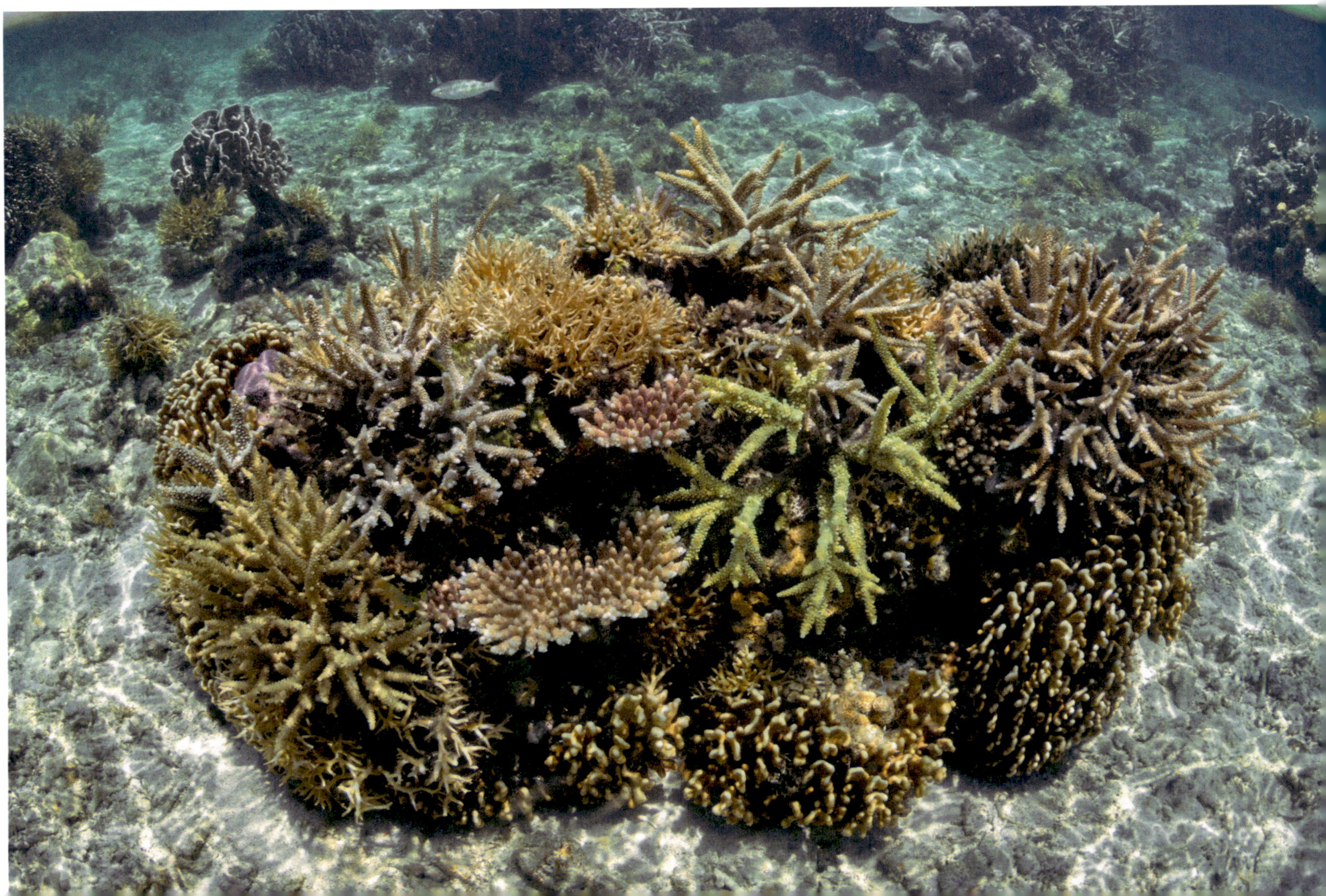

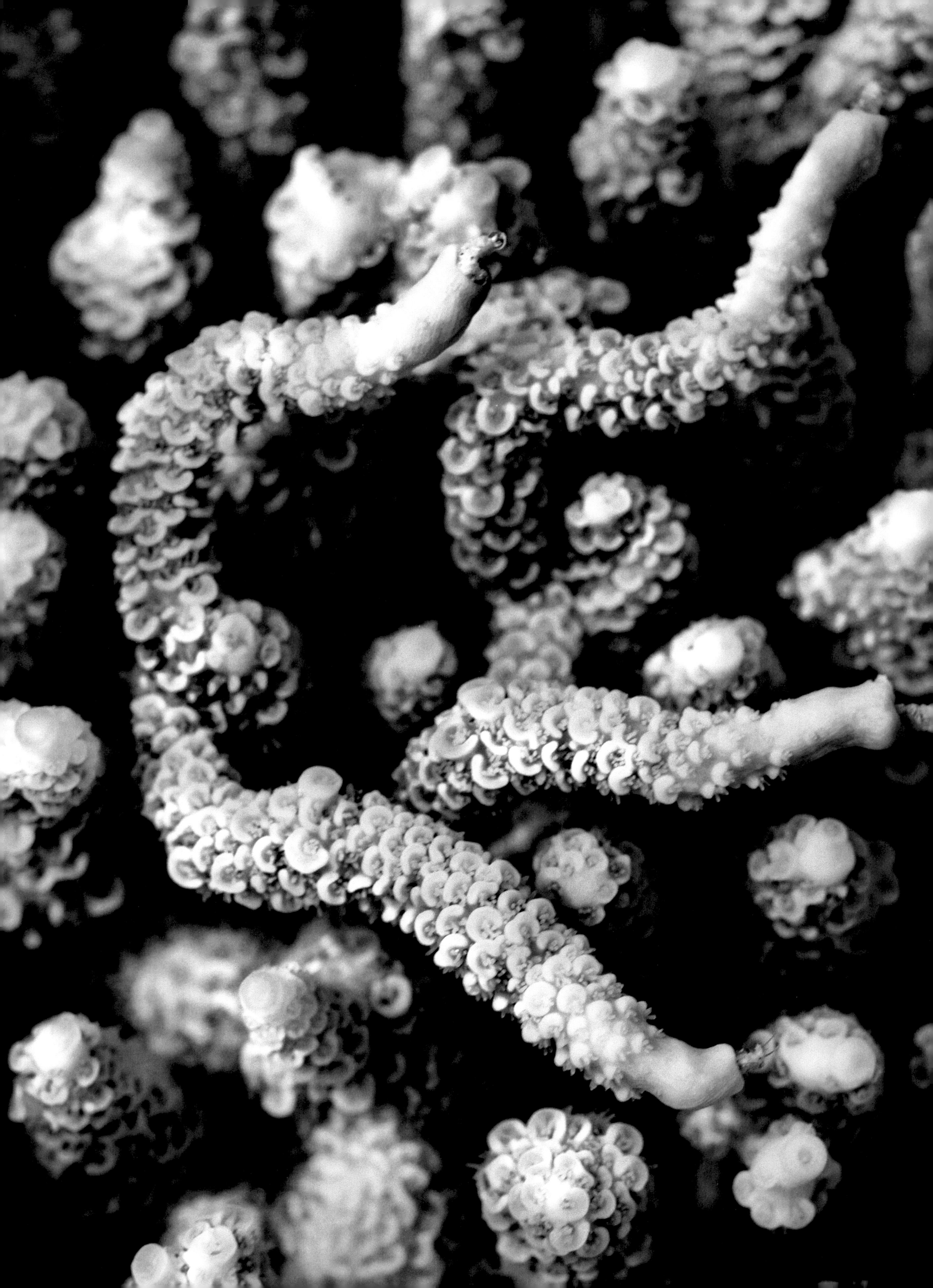

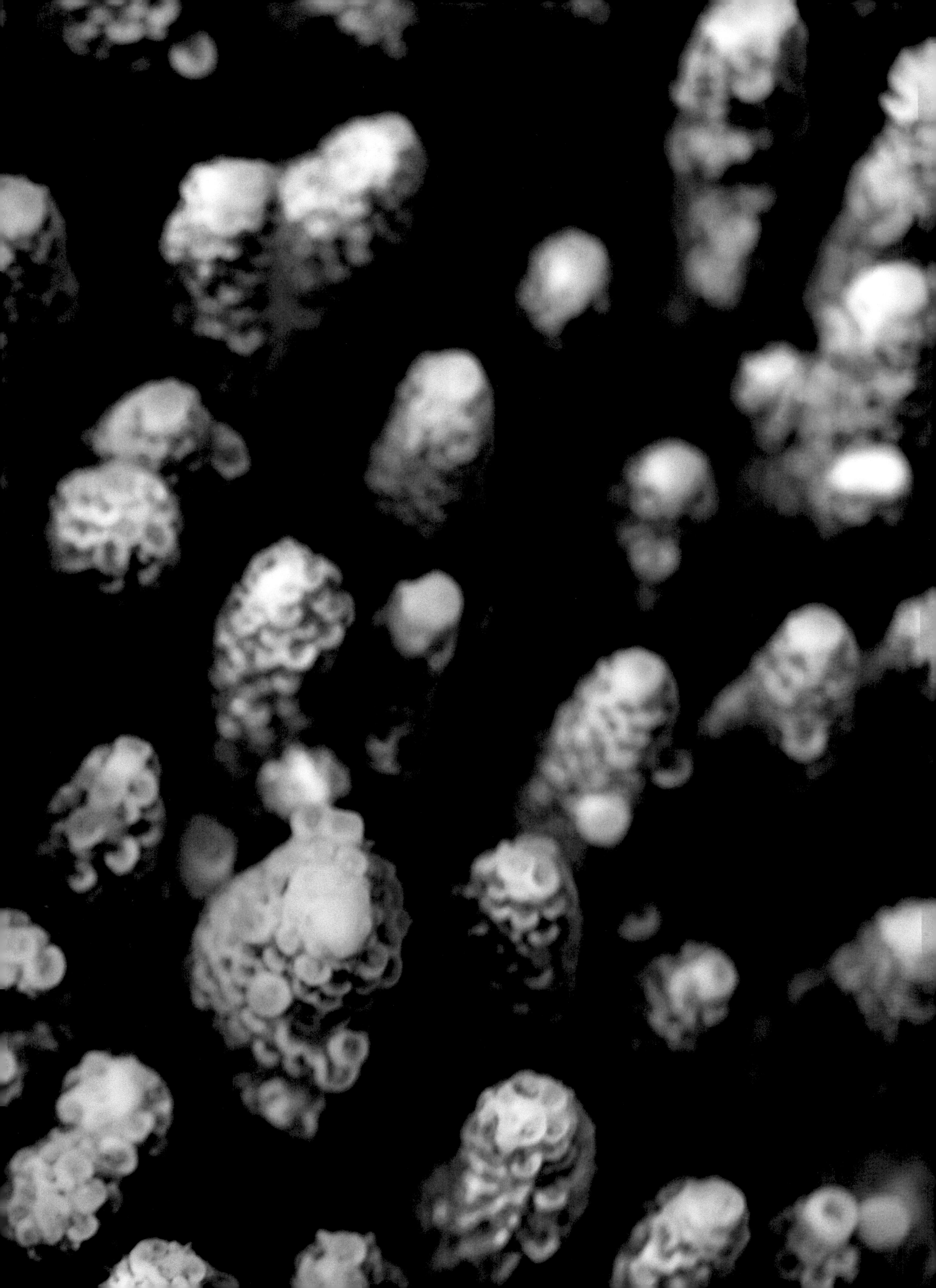

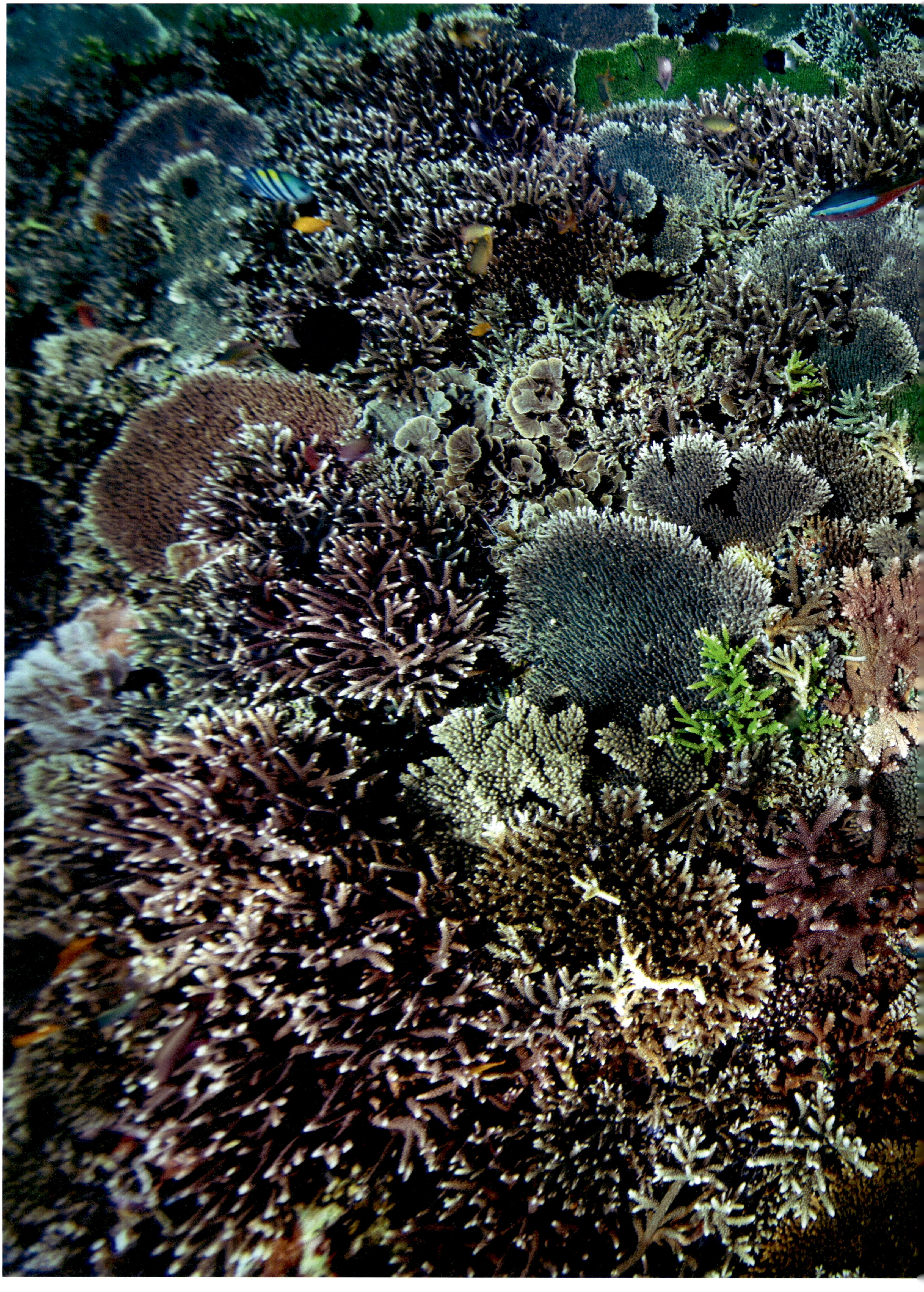

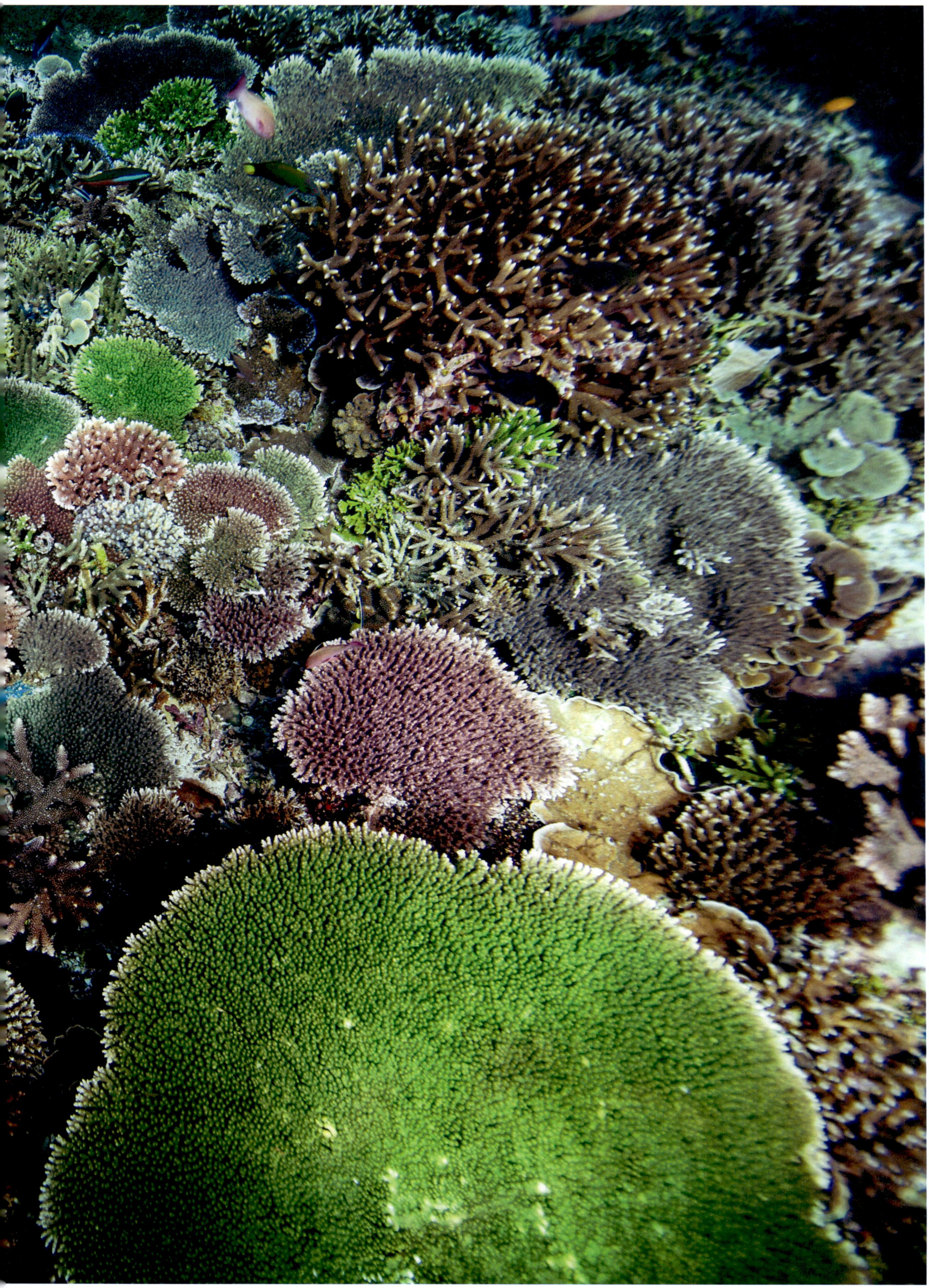

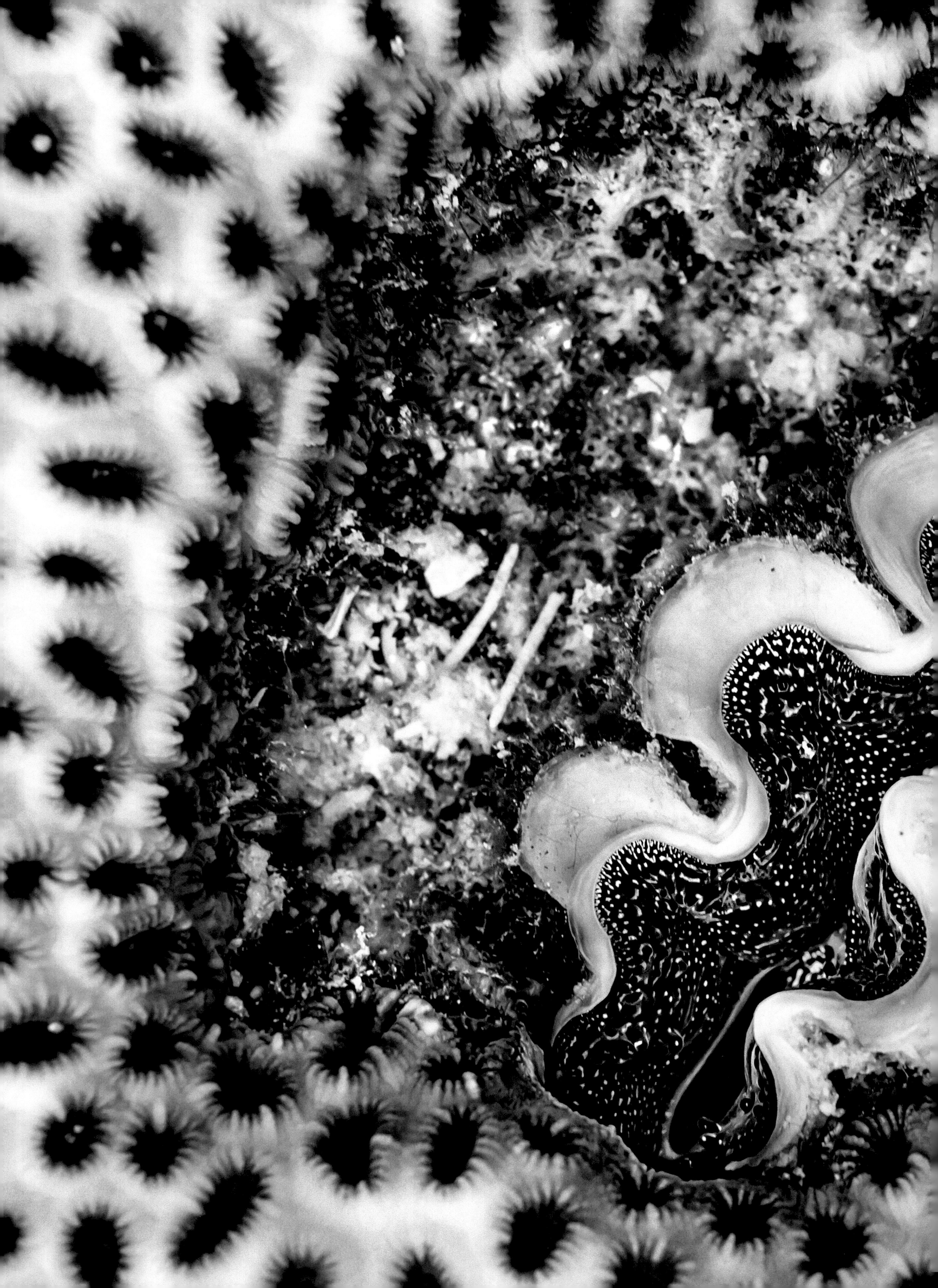

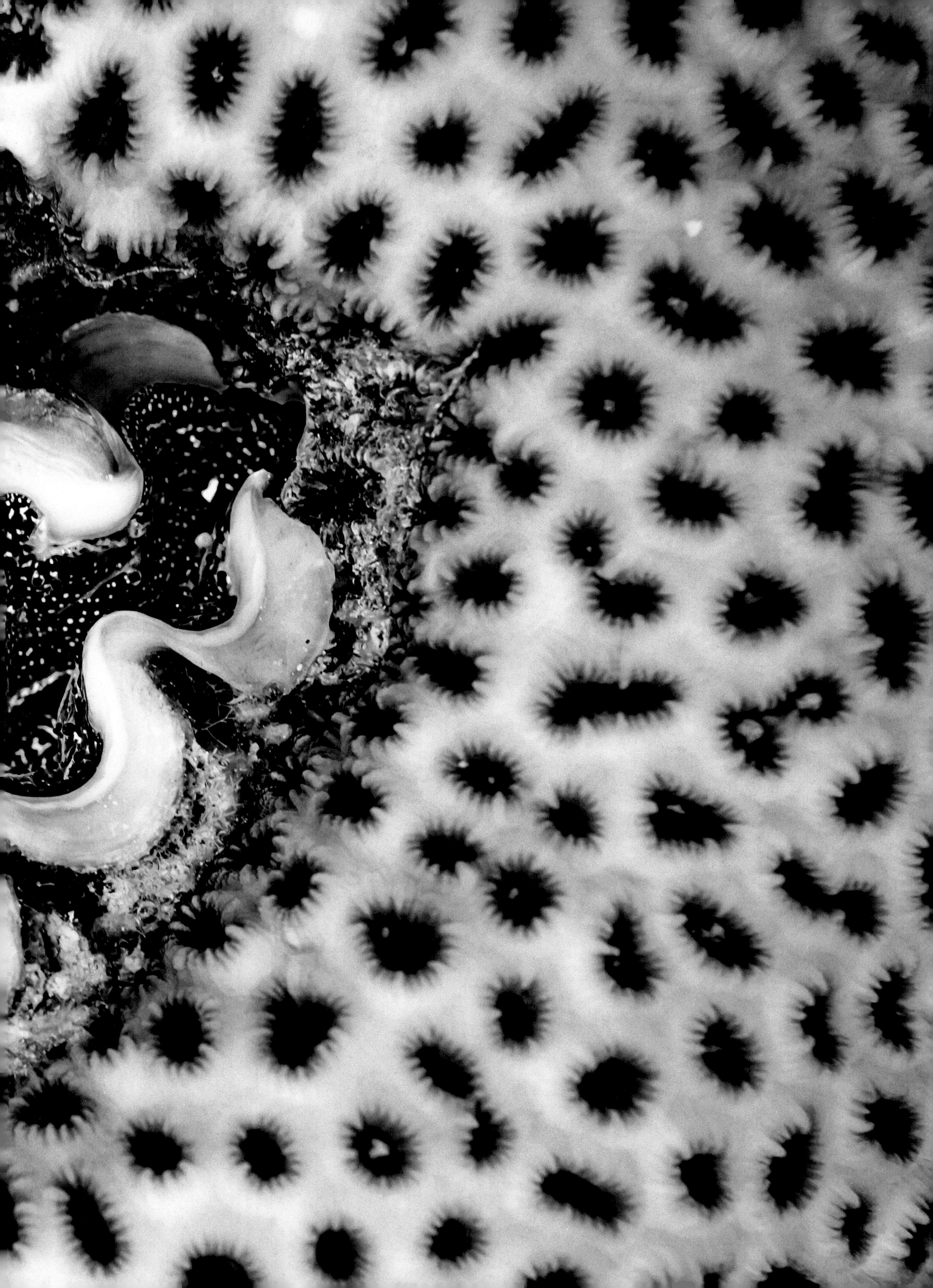

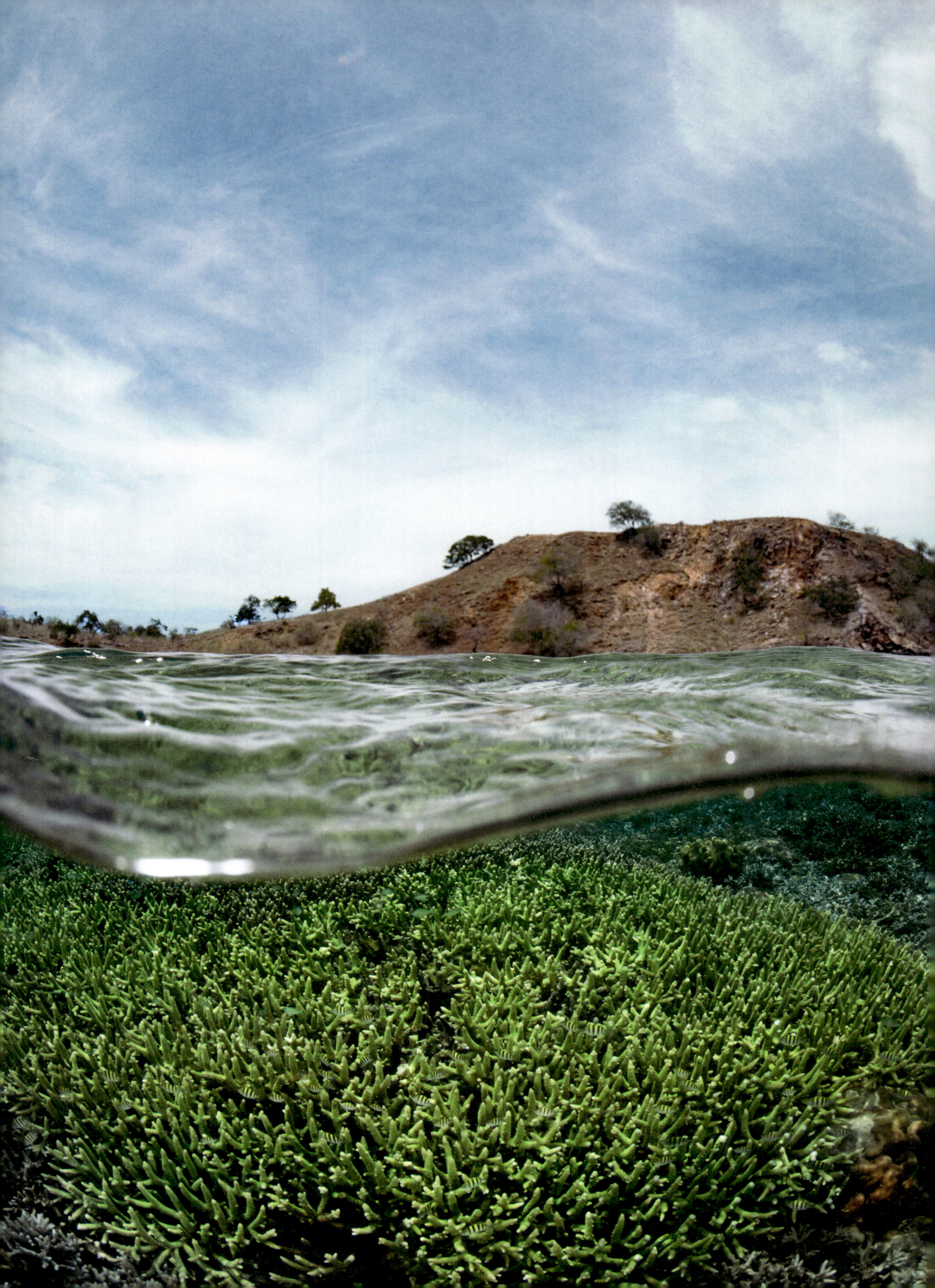

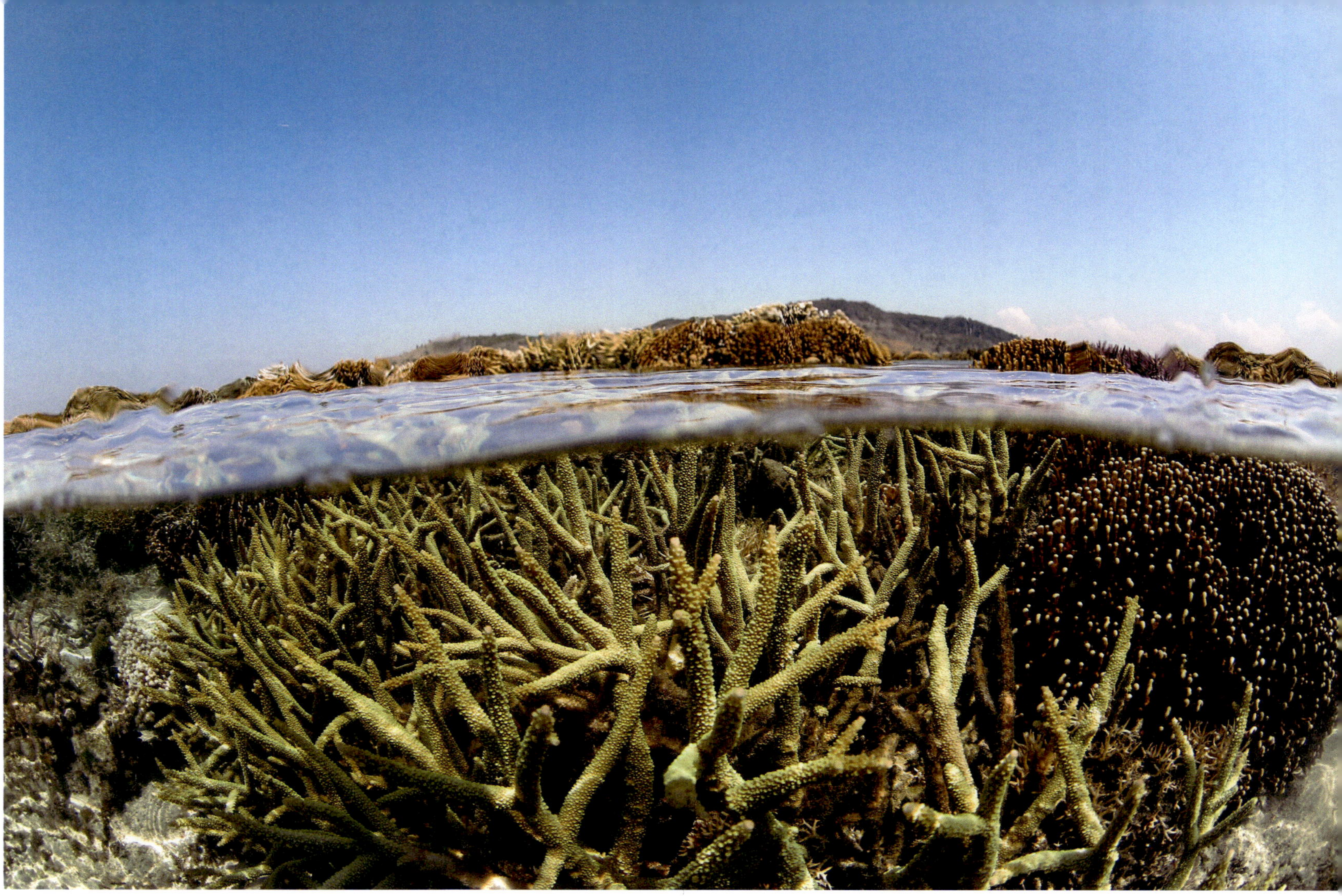

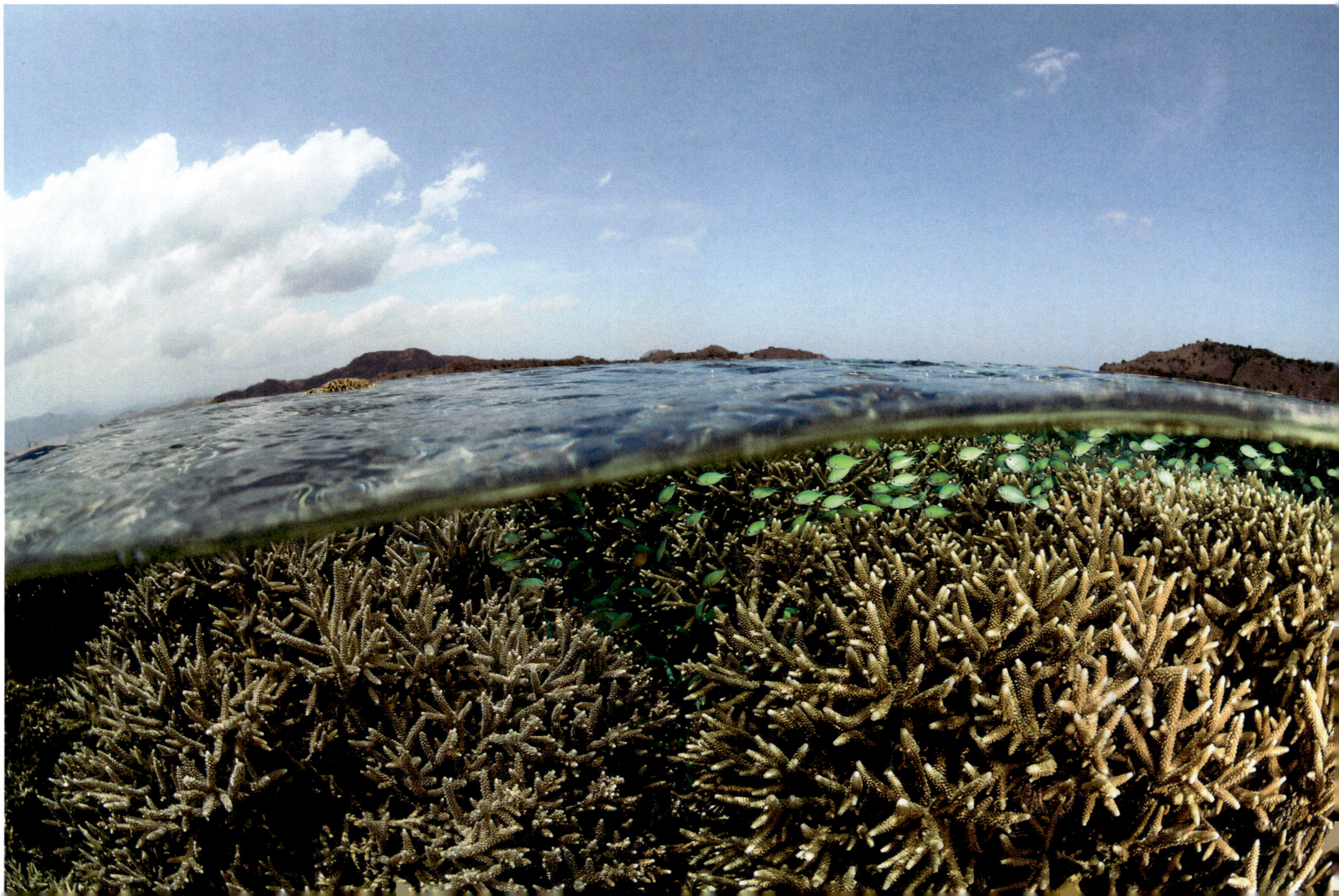

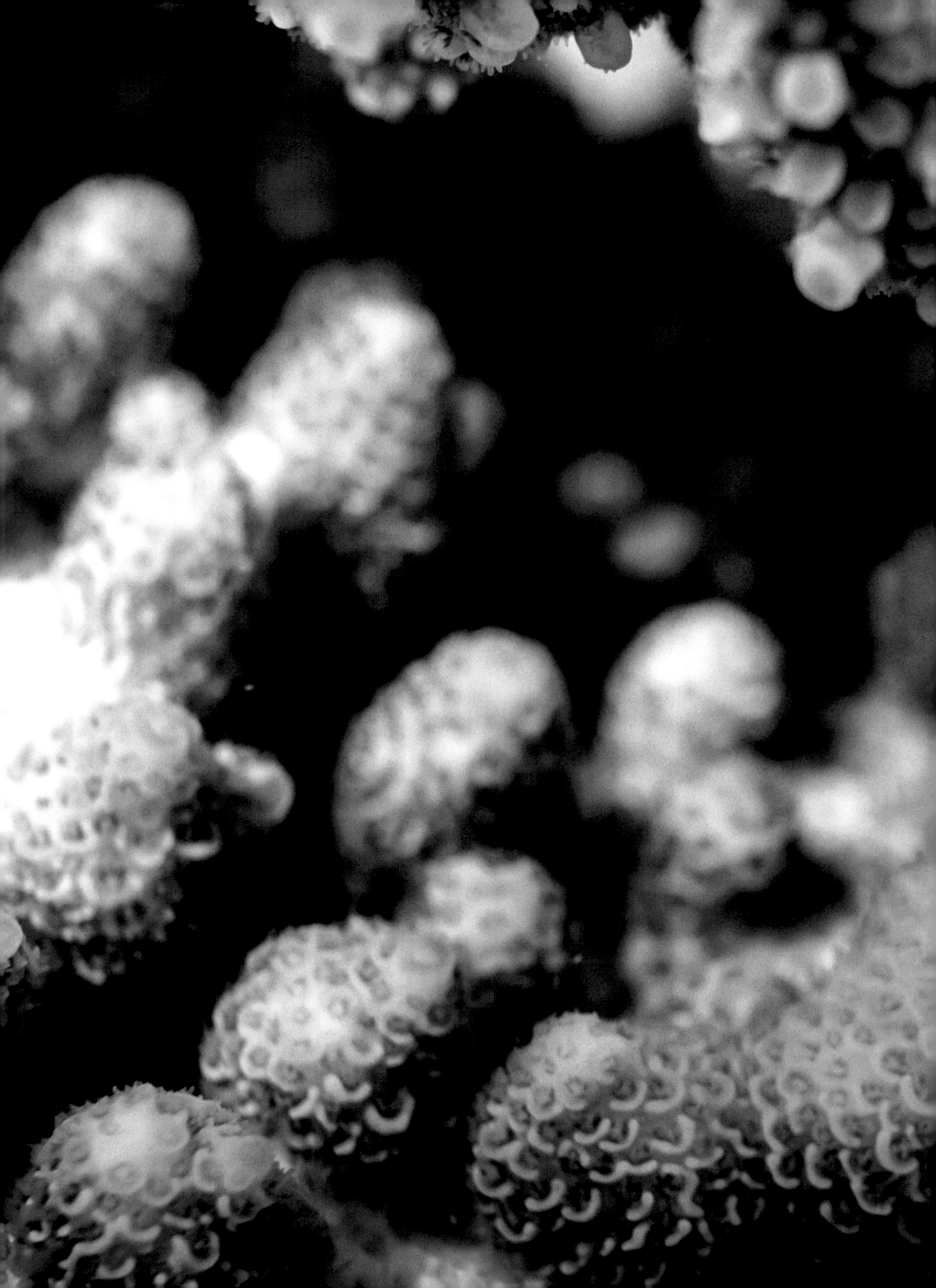

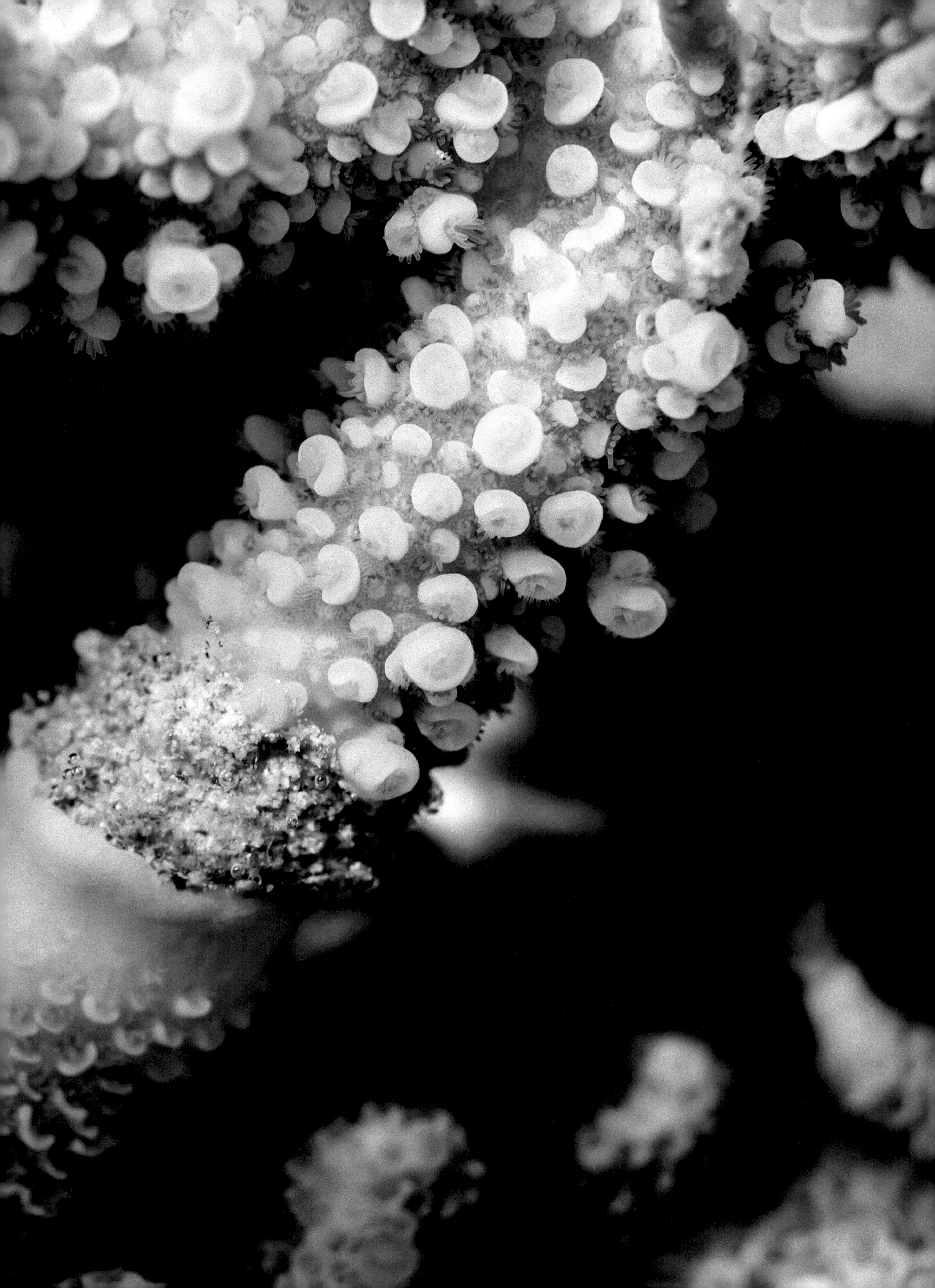

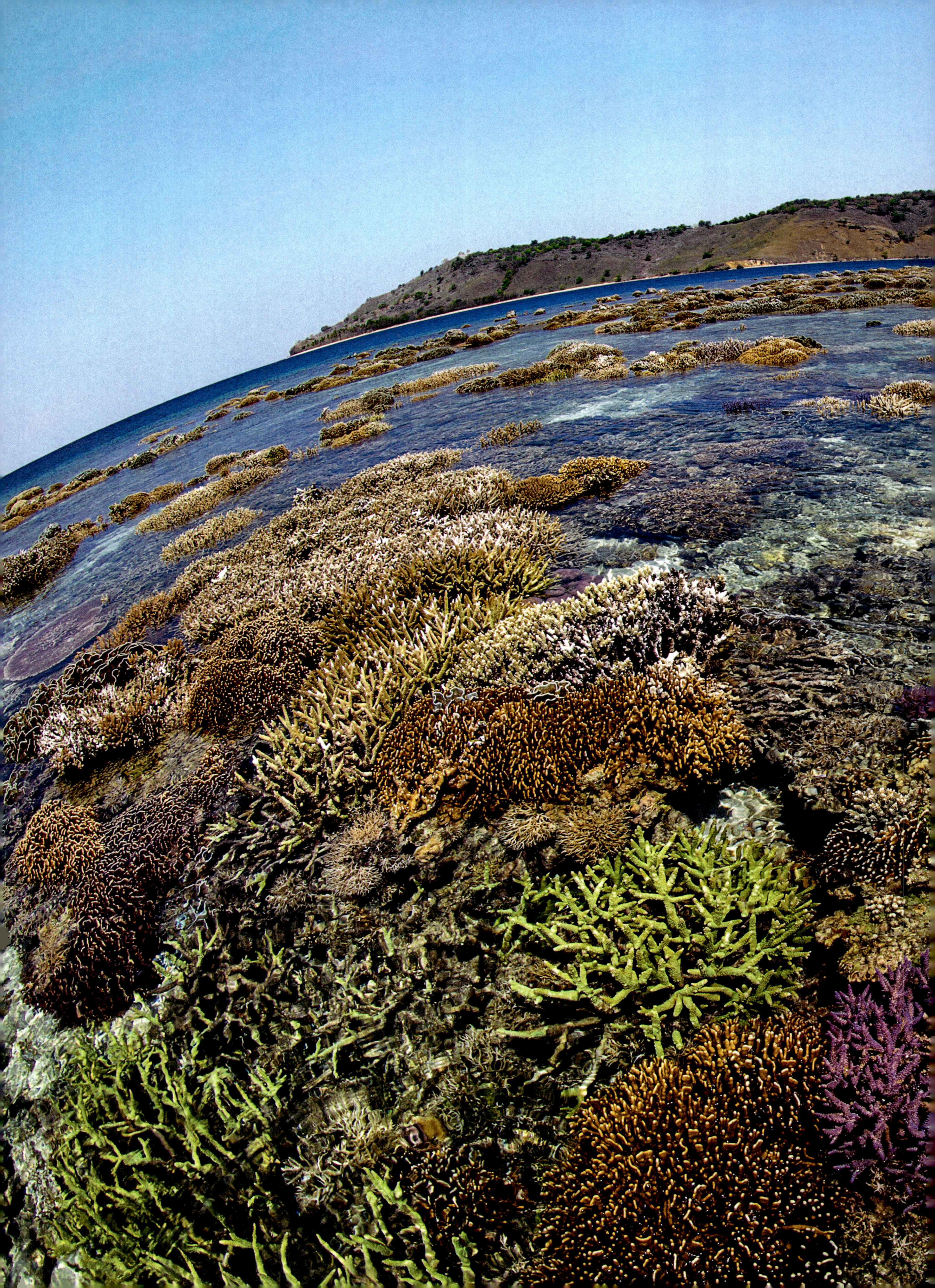

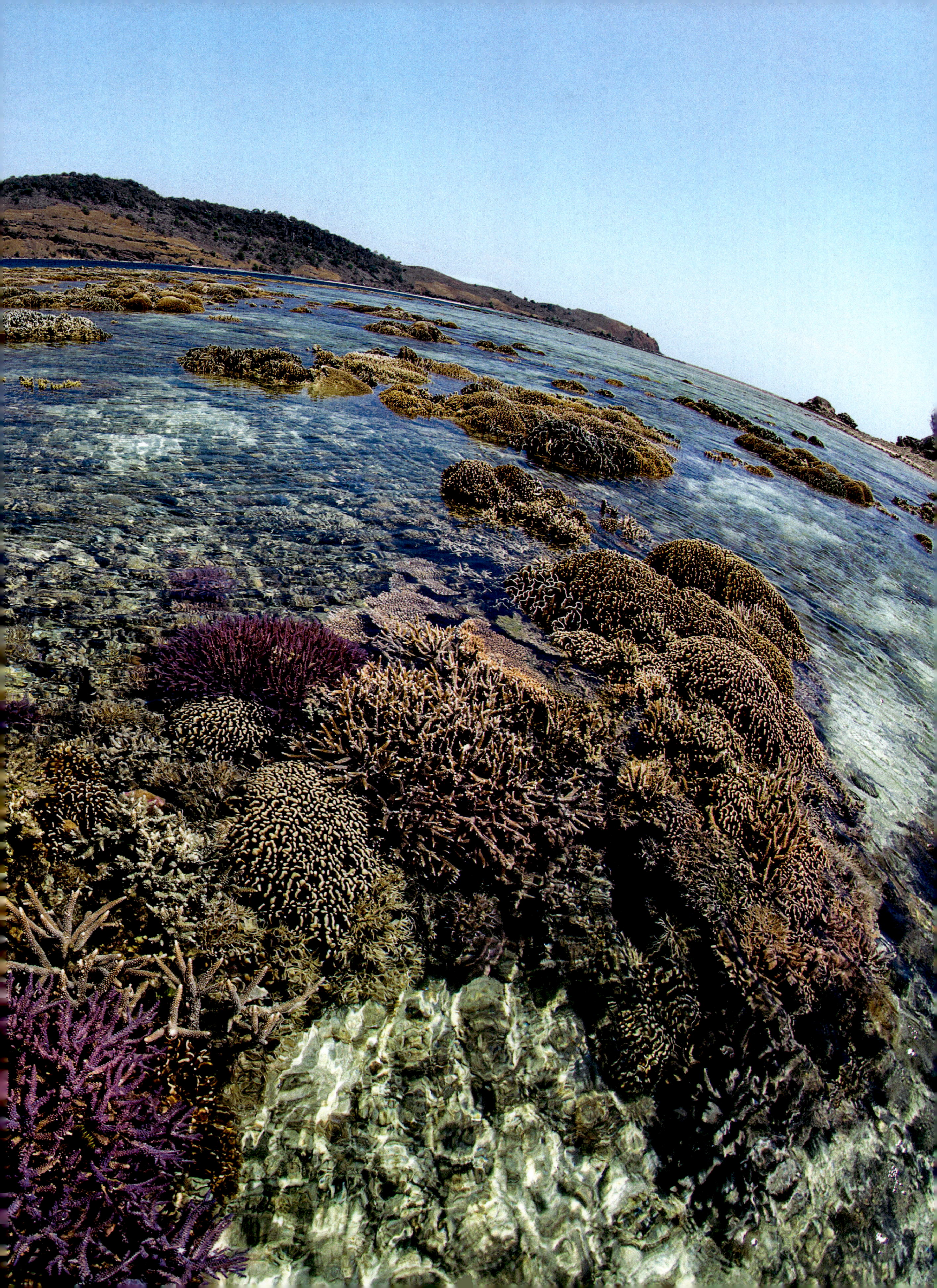

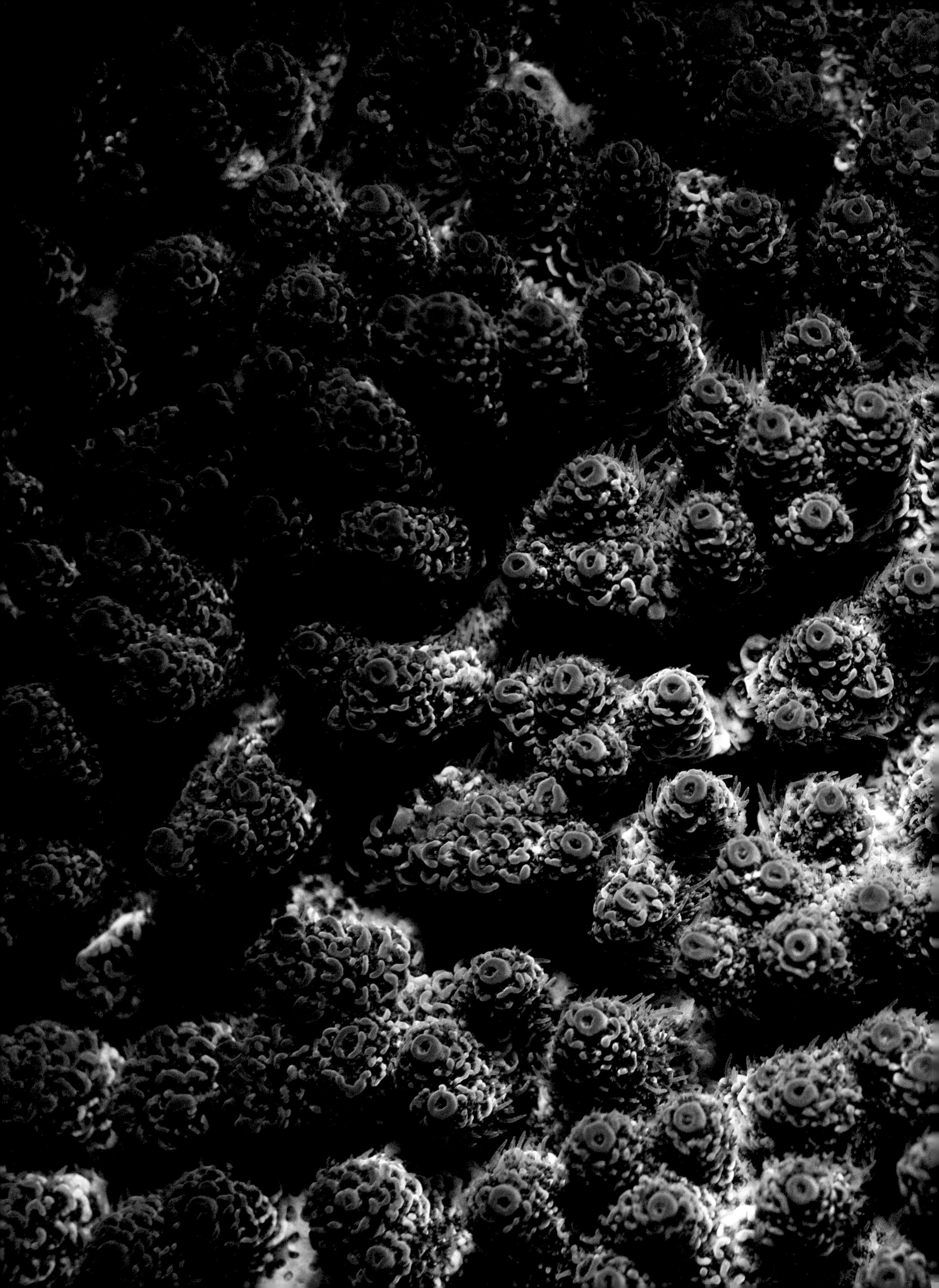

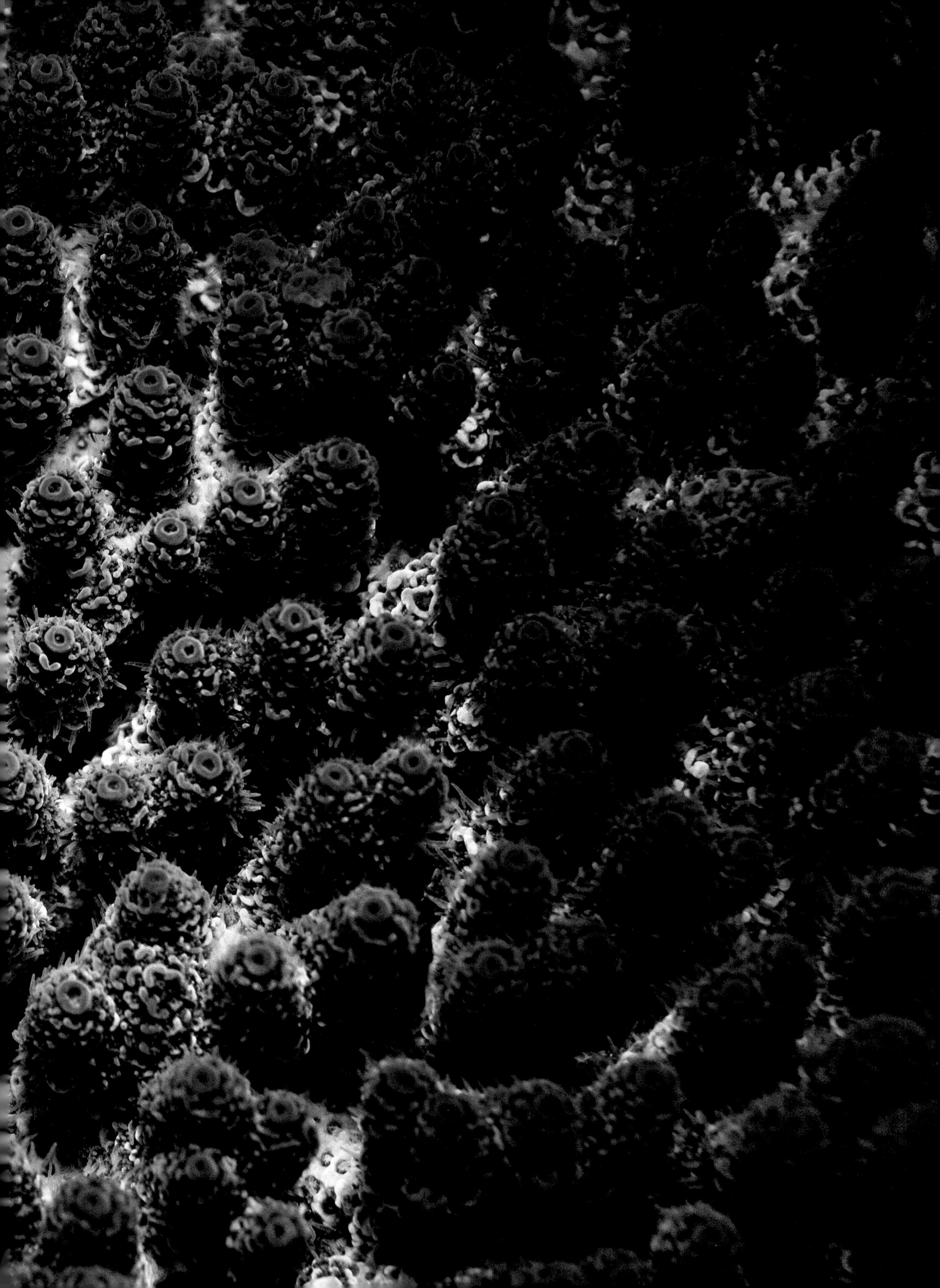

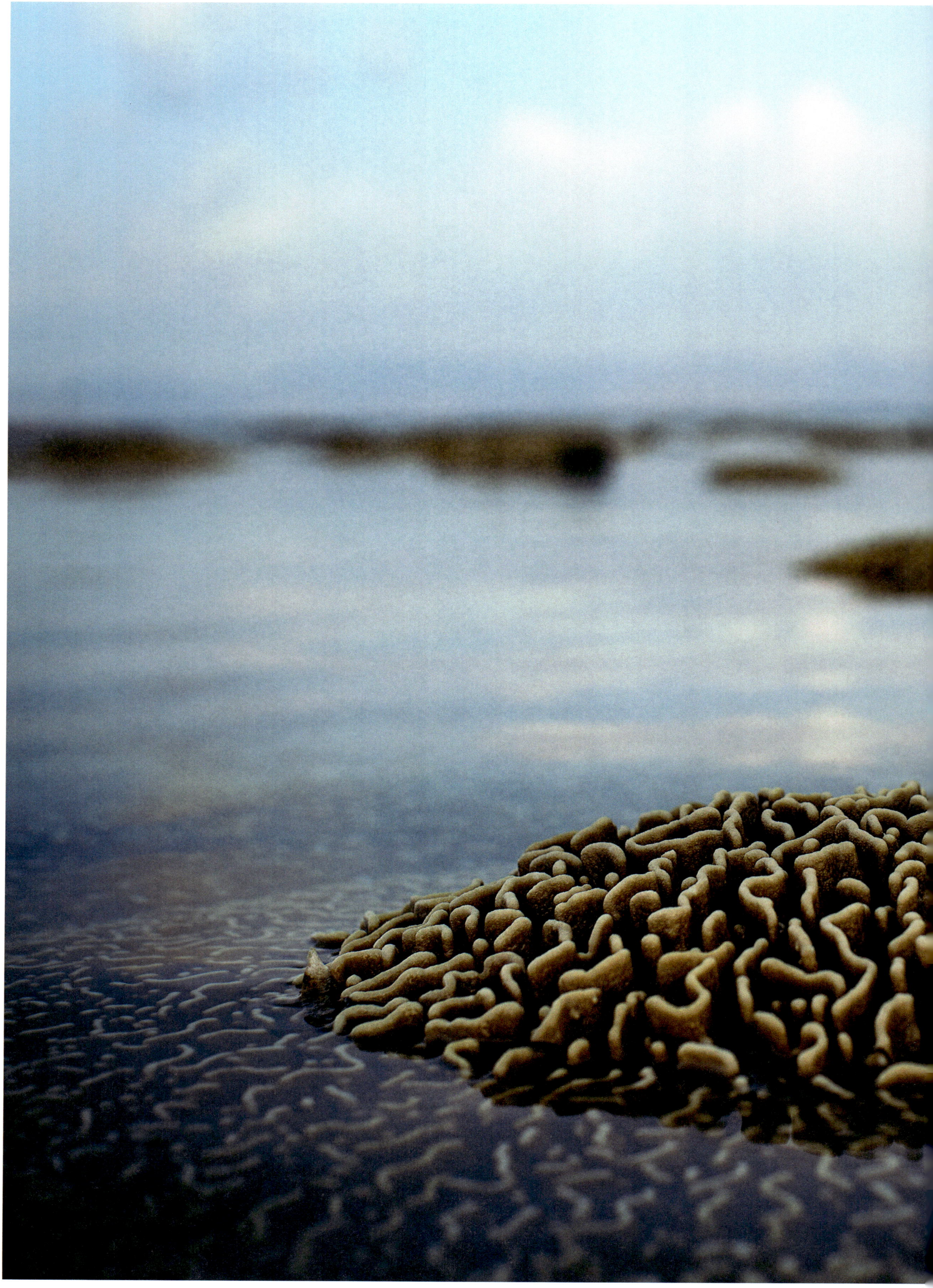

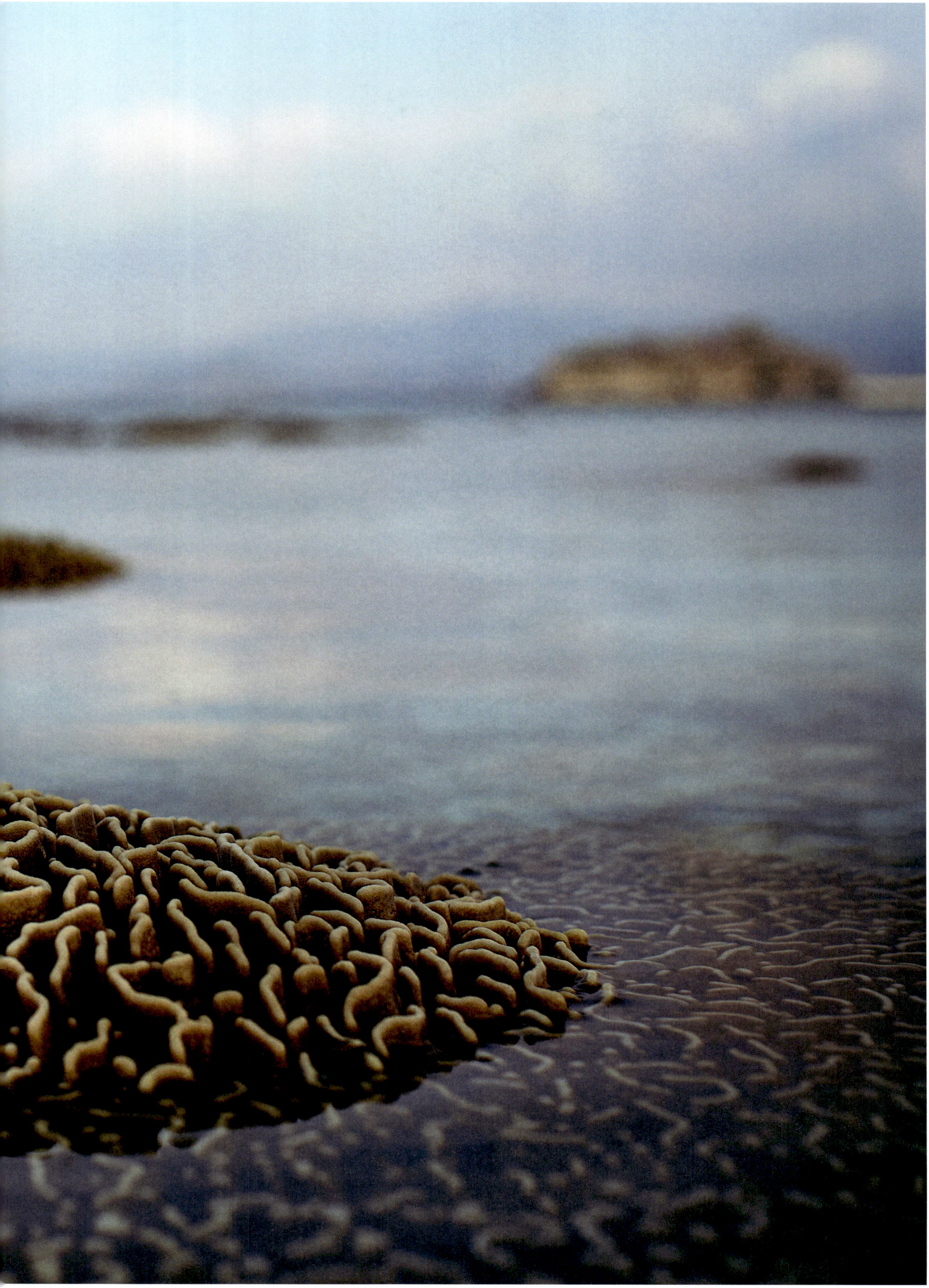

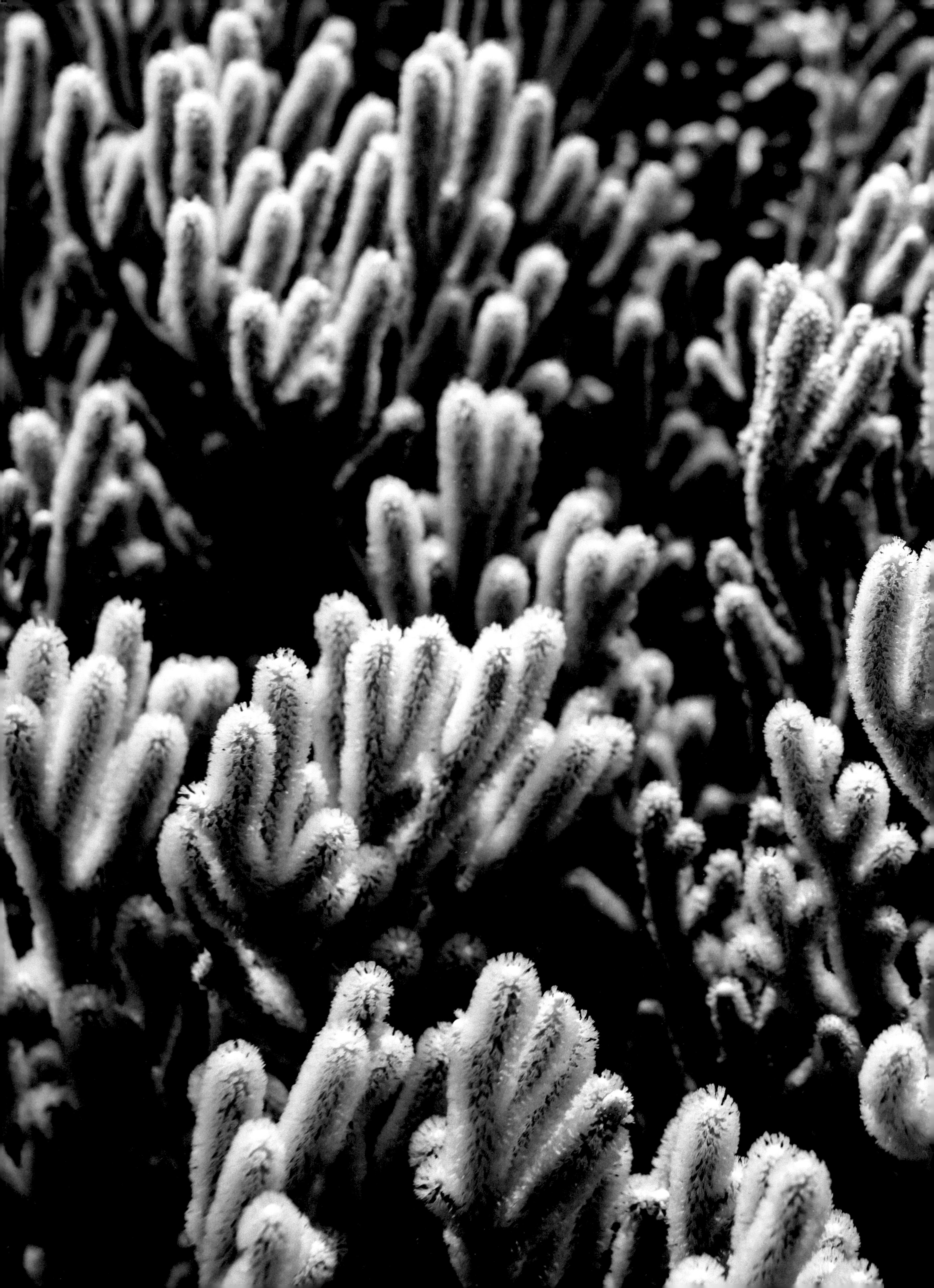

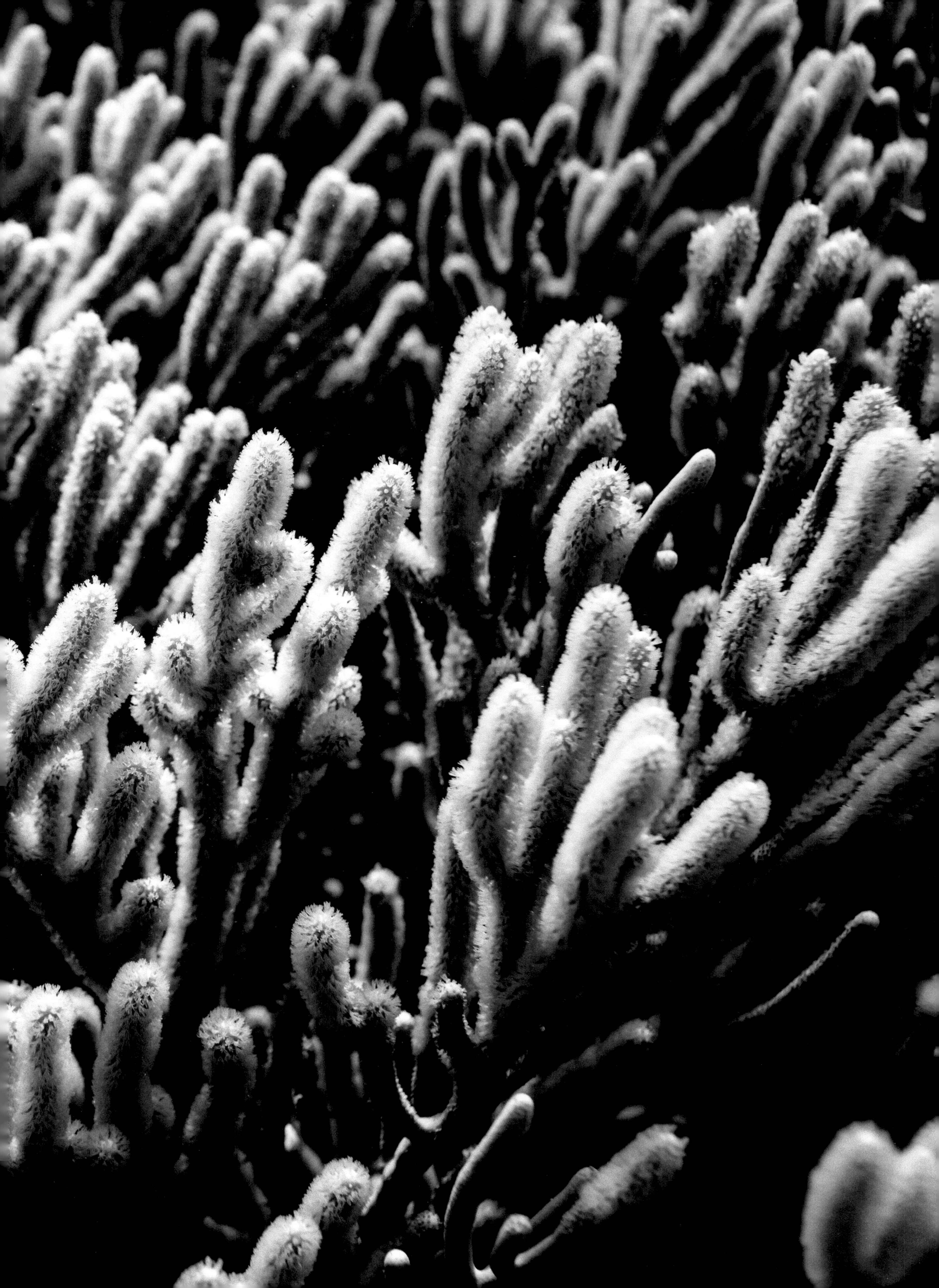

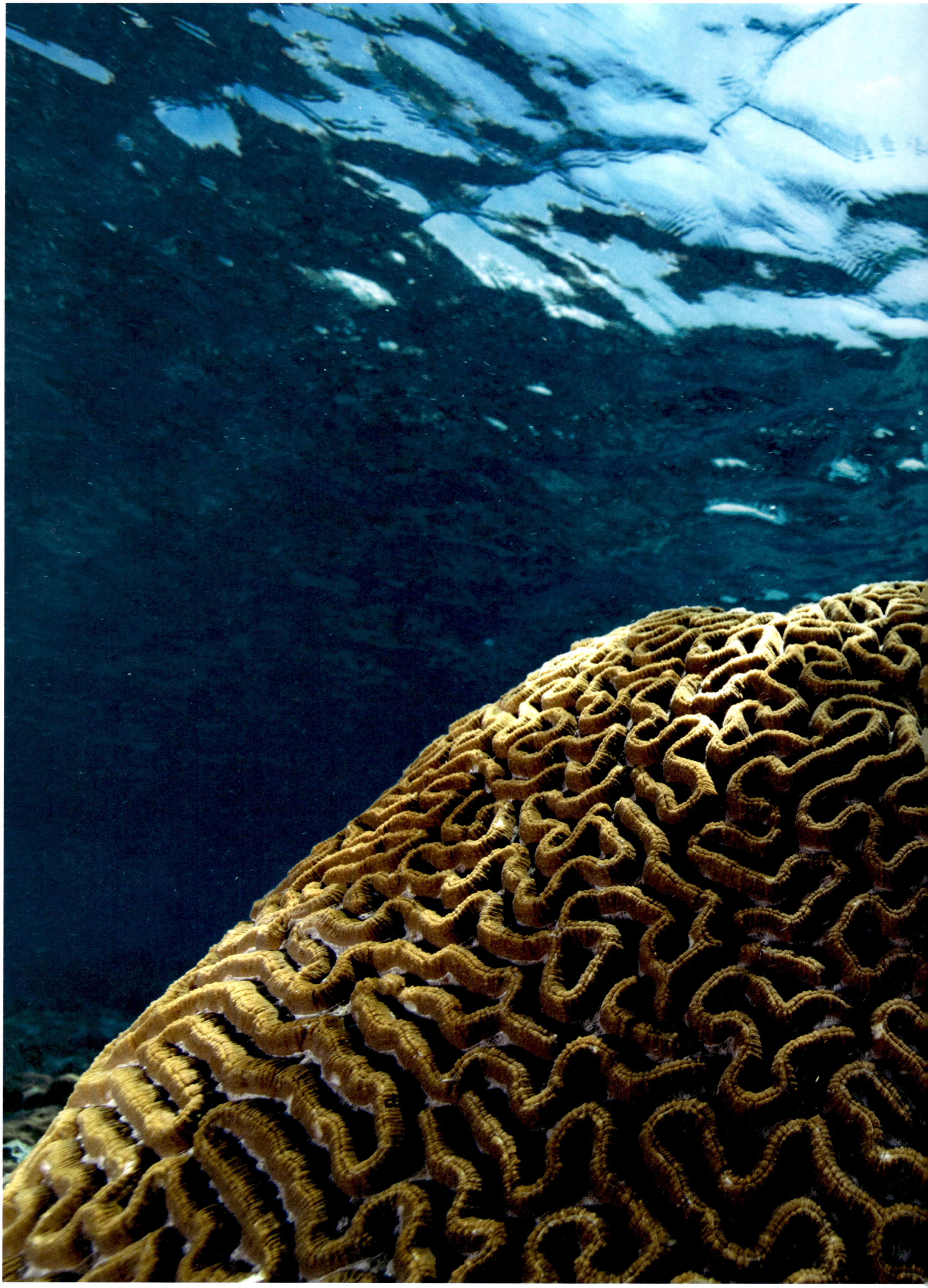

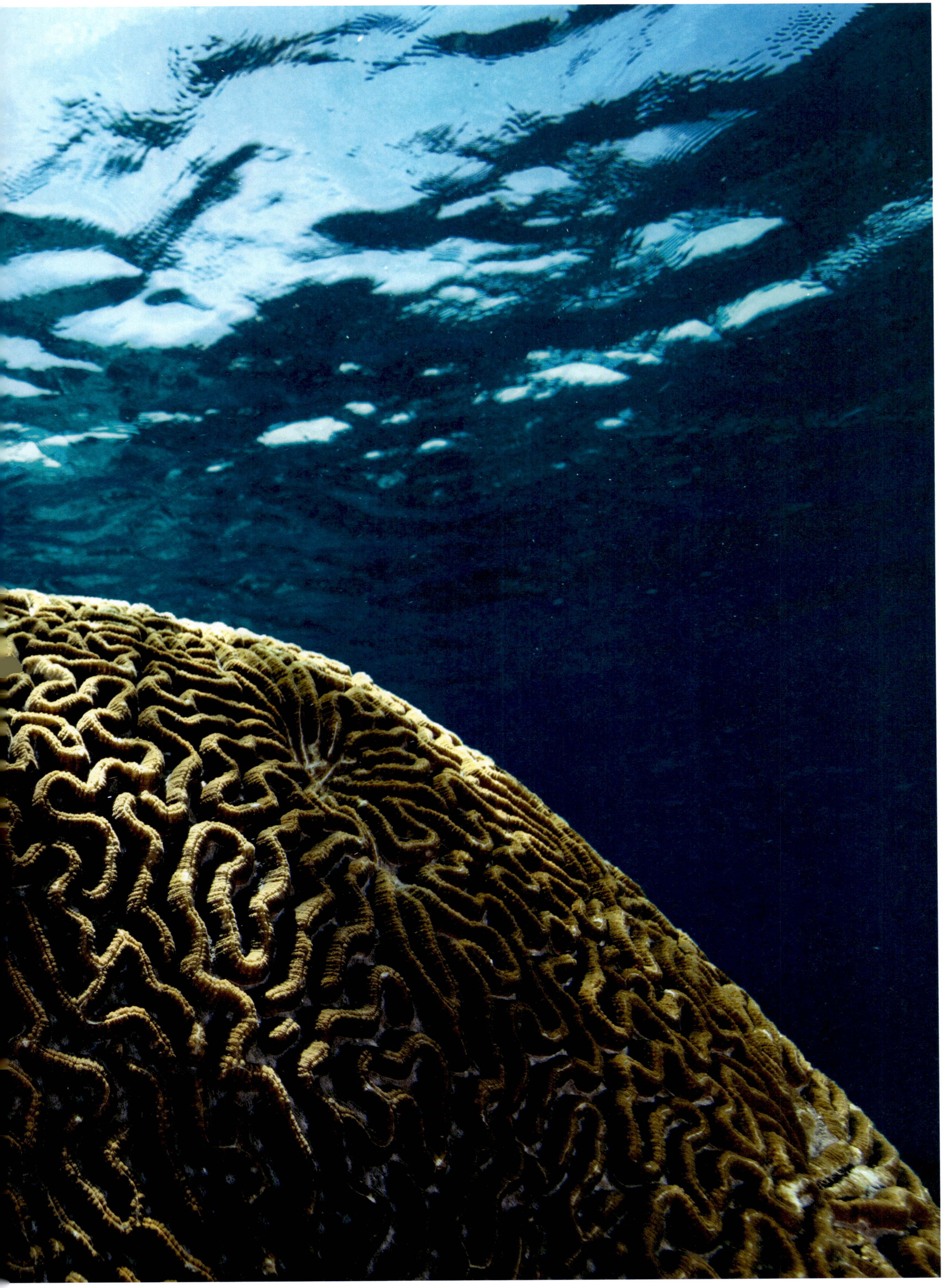

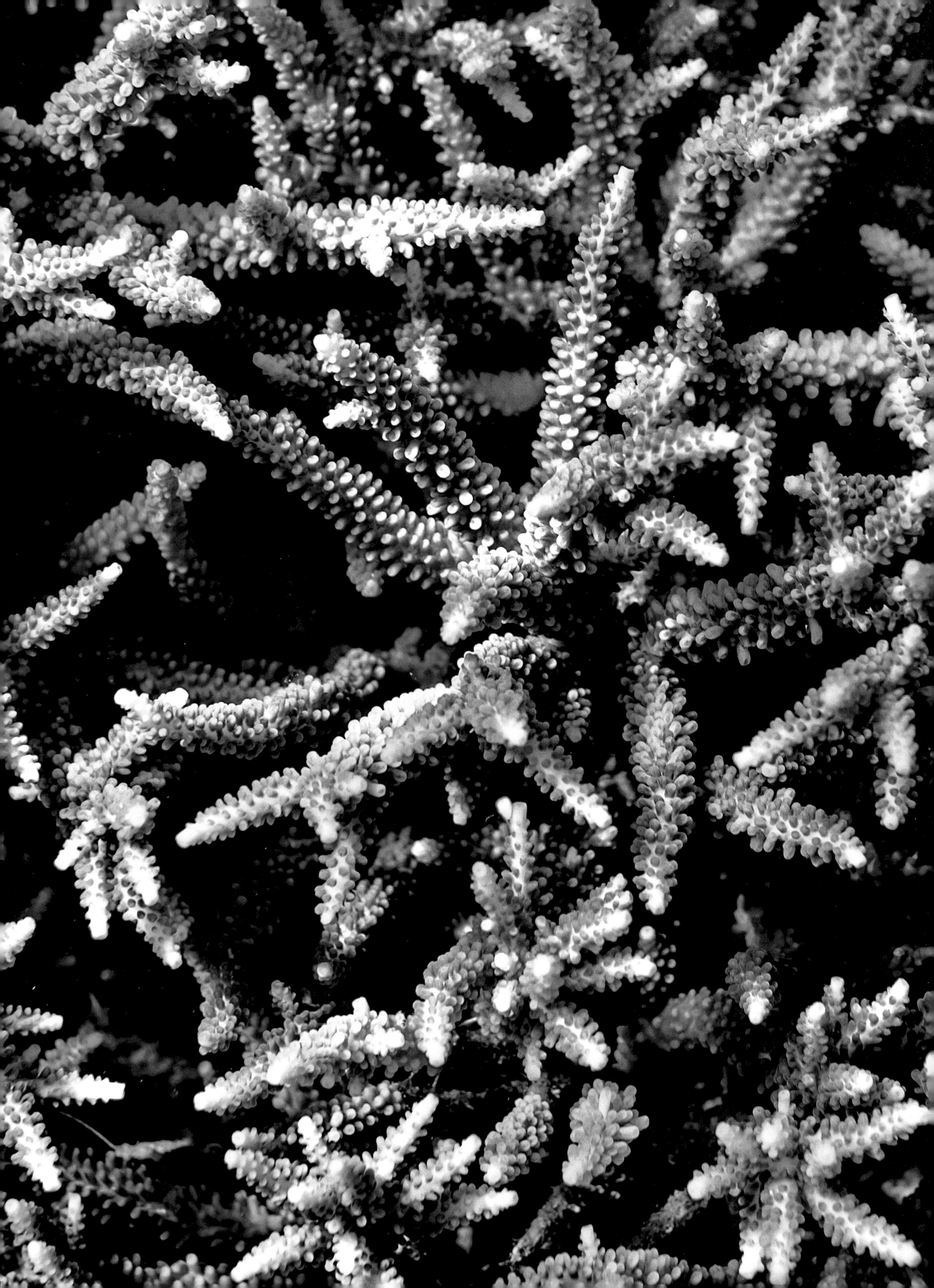

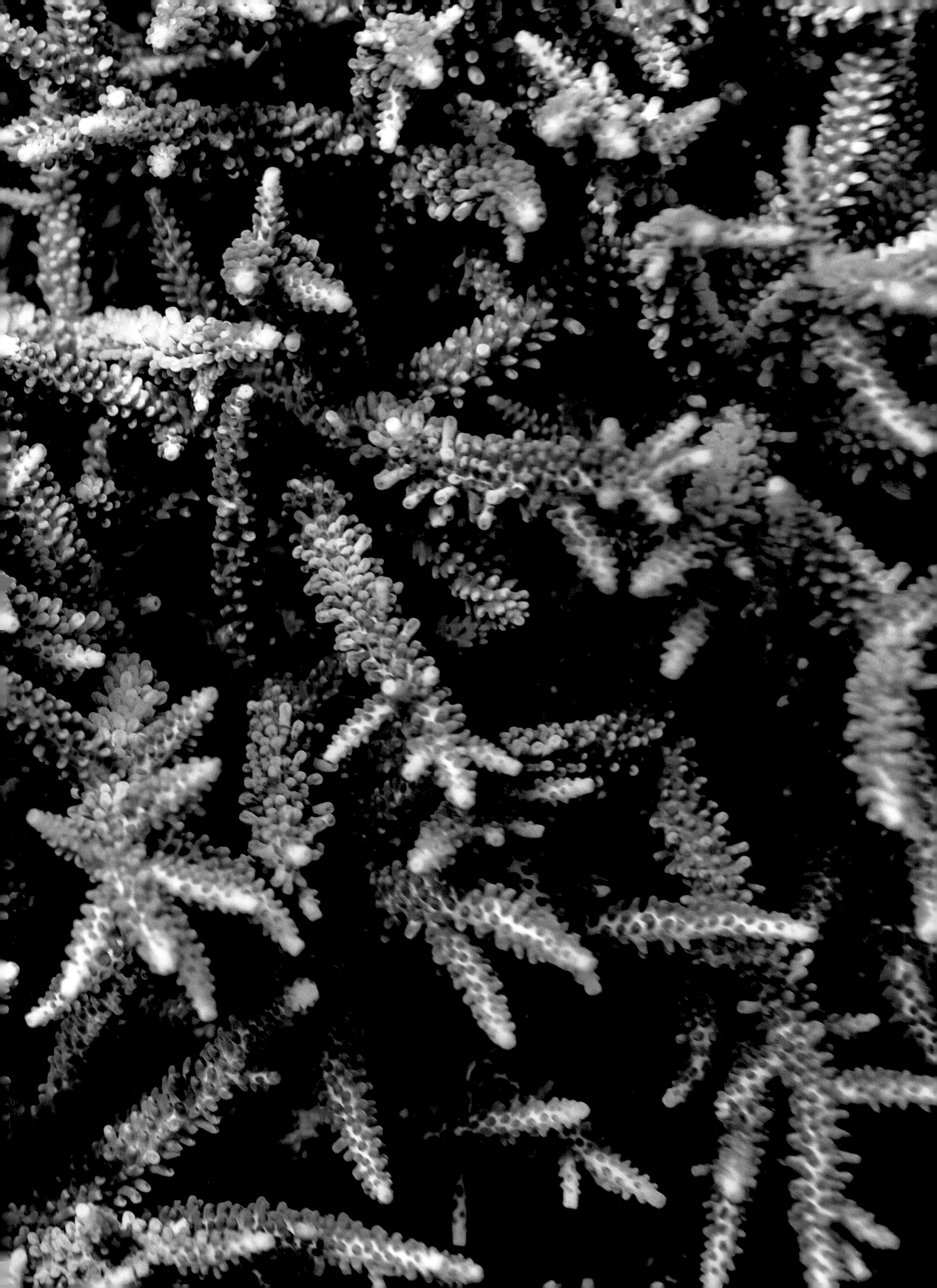

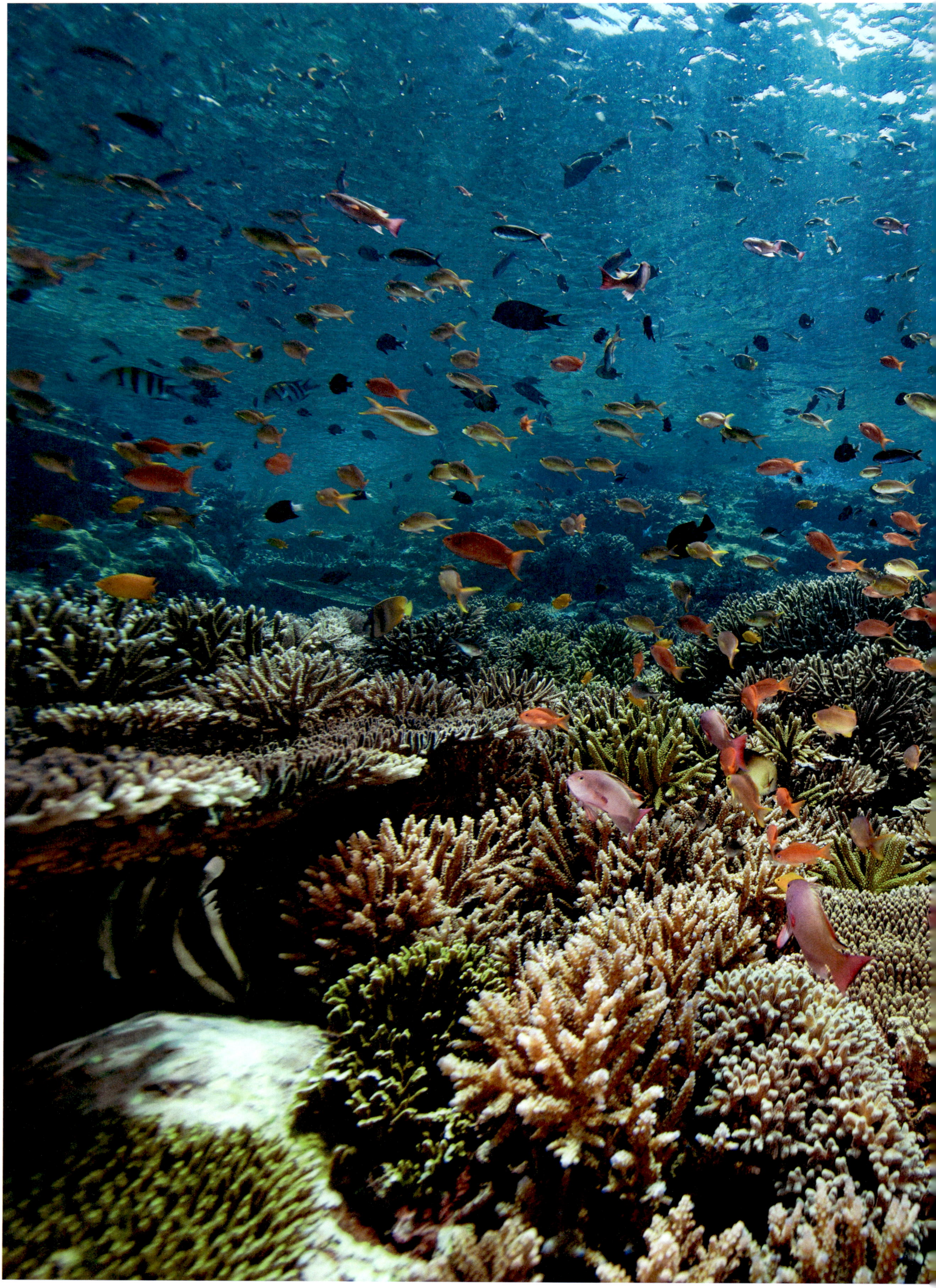

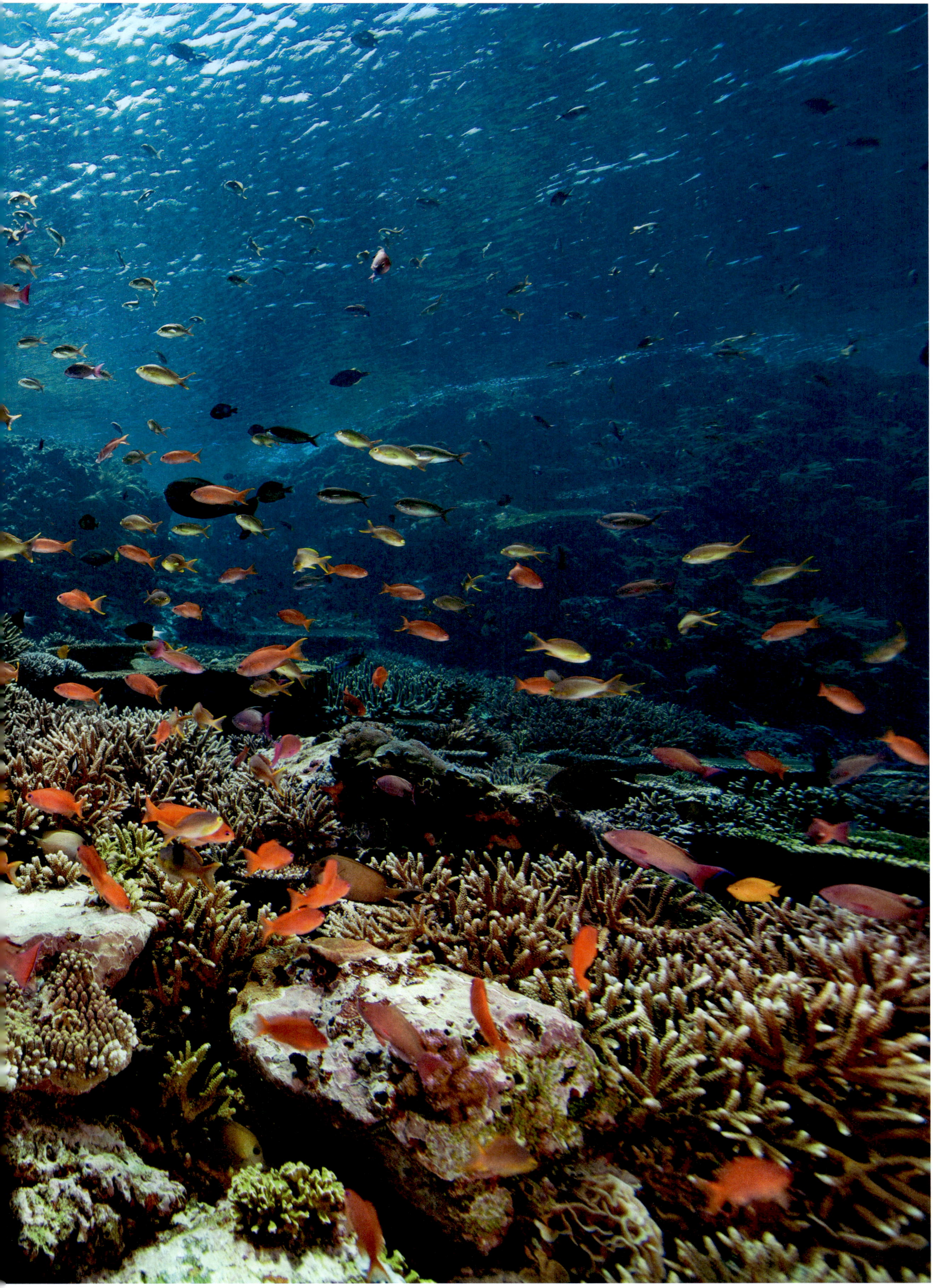

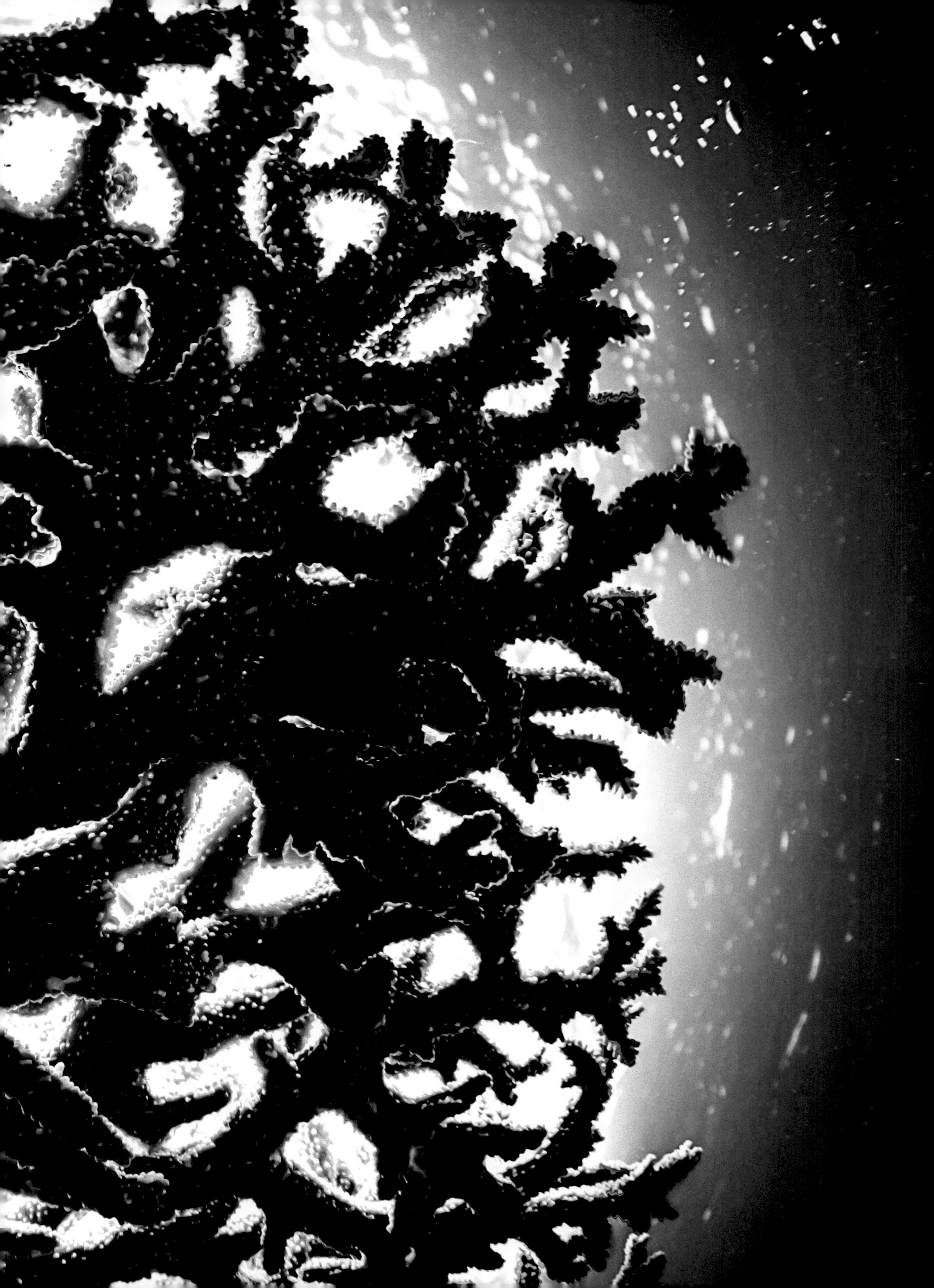

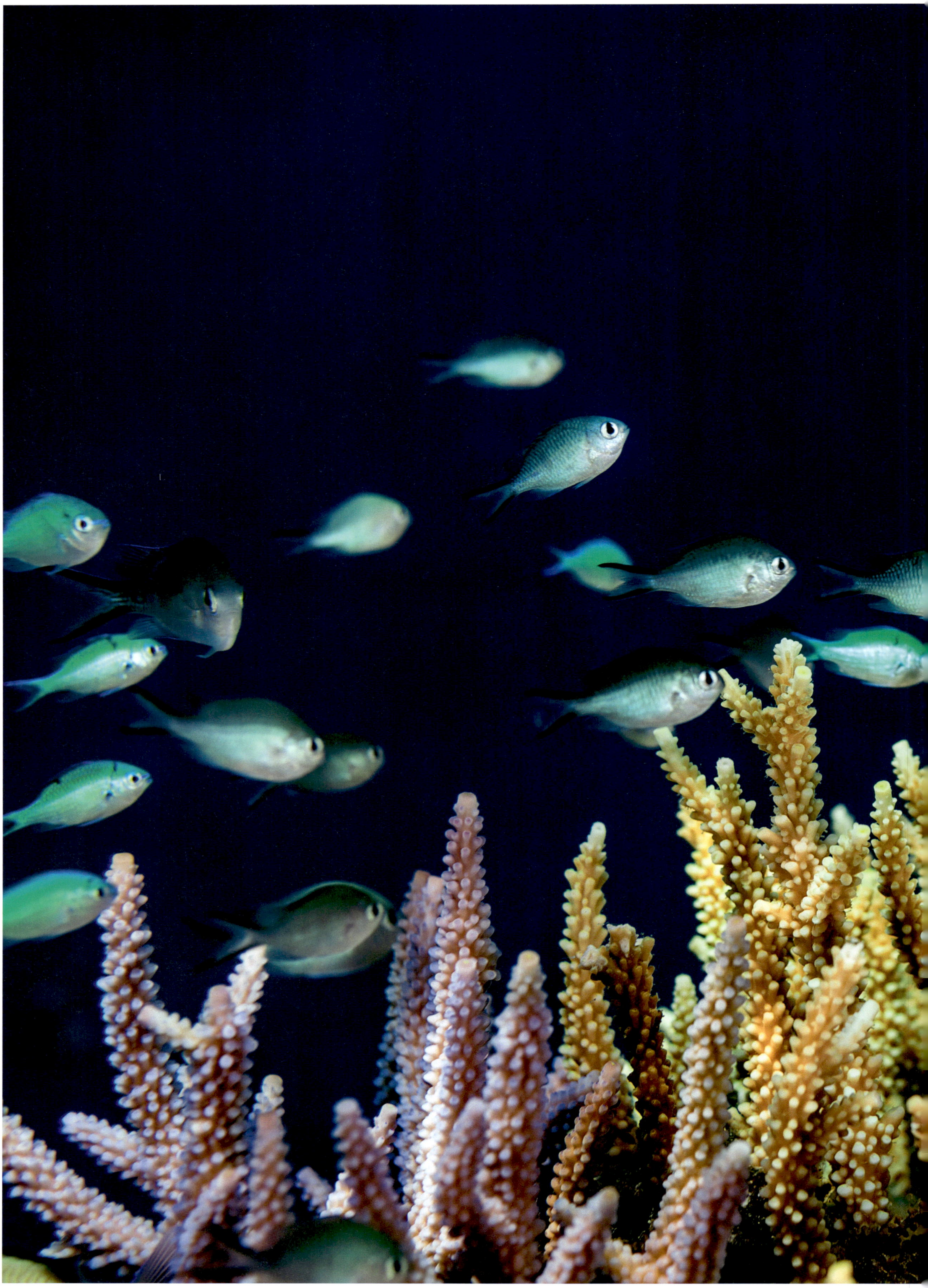

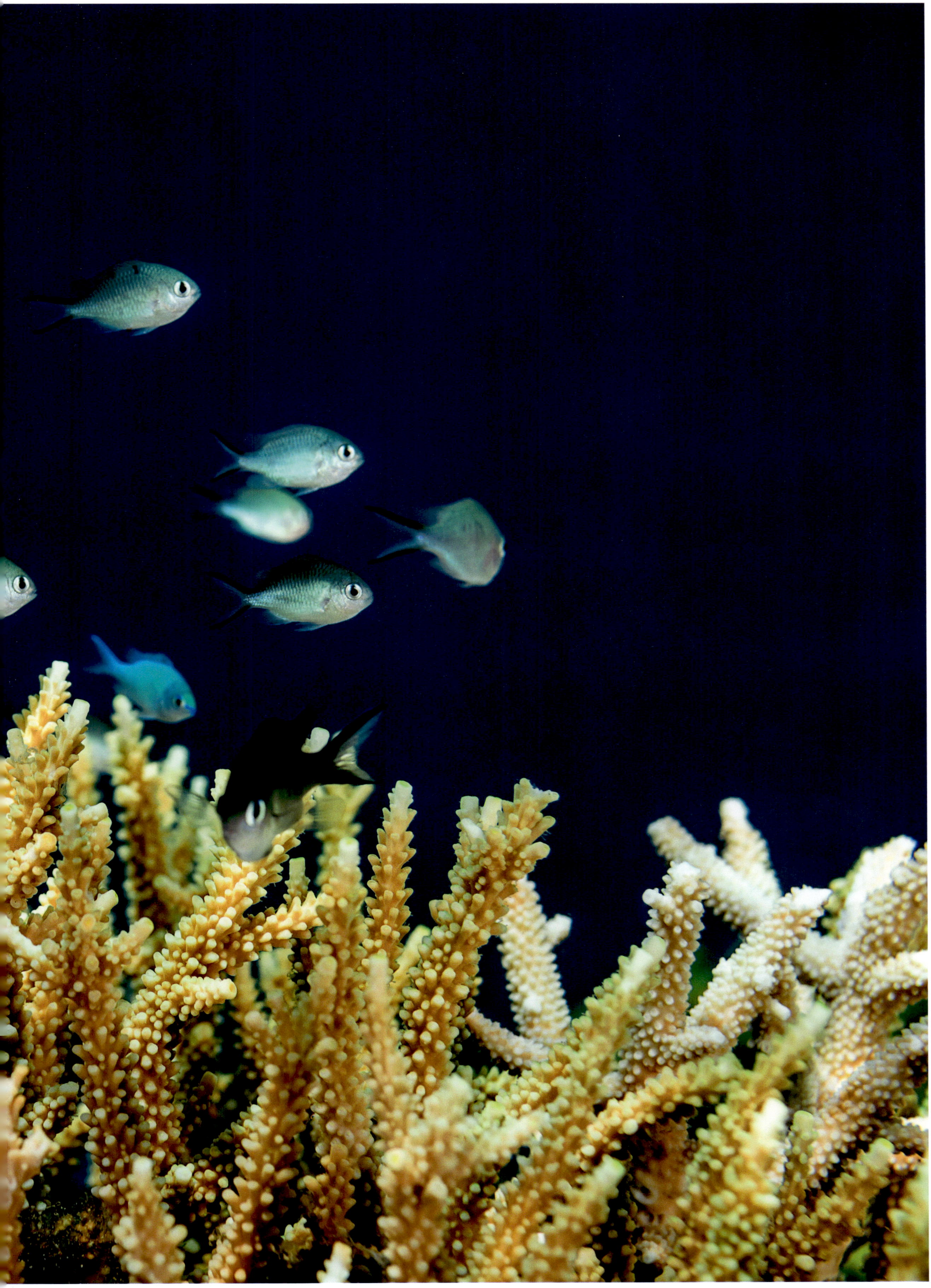

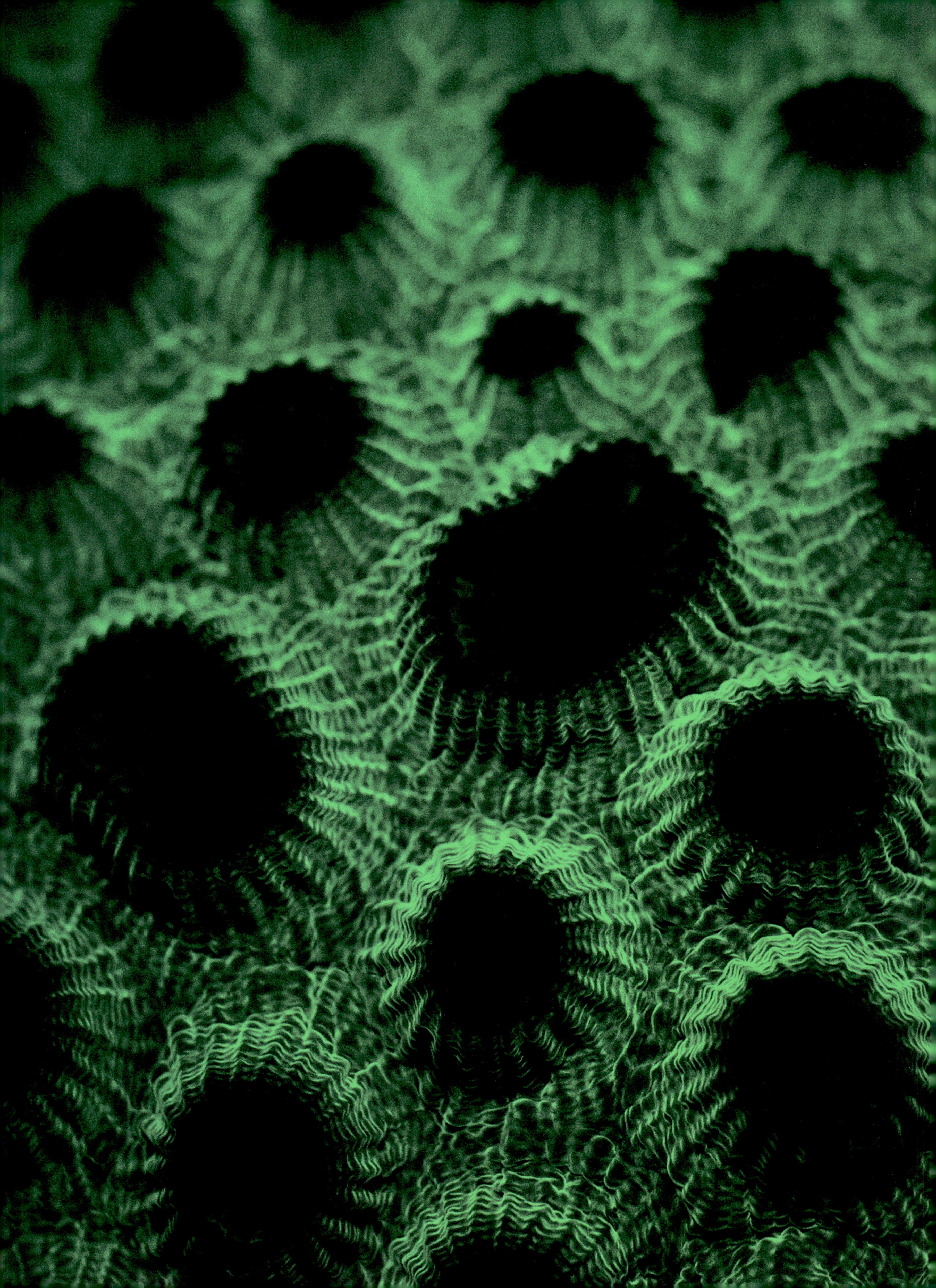

HUMAN

In these little-known places, only a few men whose livelihoods depend on the sea are involved with marine life. We find ourselves near Indonesian islands unknown to most foreigners. Somewhere in the Flores Sea, near the island that bears its name, baptised by the Portuguese for its indigenous yellow flowers – unless it was for its bouquets of coral. Seraya Besar Island is where those wandering the sea found refuge during the Second World War. They settled down and built homes on stilts on the island, allowing the island's coral reefs to rest and regenerate after human activity.

Martin decided to go to this isolated island, to share the lives of its inhabitants, to speak with them in an Indonesian whose most simple words act as a bridge between the various dialects across the 17,000 islands of the archipelago. As he earns their trust, the foreigner gradually disappears, he becomes part of their quotidian, goes out on fishing expeditions, becomes friends with Komang the fisherman, or Mirah, the little girl who follows his photography trips around the village.

Martin captures a way of life that is dependent on the sea and fishing, beating to the rhythm of the seasons and the changing climate. He conjures up the daily life of a village without roads or running water and with only unreliable access to electricity. He grasps the hidden meanings of facial expressions, fishermen's gestures, the way that children play or help out around the village. He shows the men's feet, hard like the baked earth, just as he shows their ability to swim deep and propel themselves to several metres' depth.

Those nomads who put down roots on Seraya Besar tried to tame the land, to shield themselves from nature's hostility. They built some houses from concrete, which is perhaps more comfortable, even if the traditional houses on stilts are more efficient at letting air circulate and protecting them from dangerous animals. They built a sea wall to protect them from the rising waters and storms. Some mangrove patches were cut down, taking away the beach's natural protection. Wherever concrete was placed, life disappeared – forming, in the eyes of the photographer, chaotic landscapes.

Around Seraya Besar, years of blast fishing have mutilated the coral reefs and sometimes injured people as well. Fishermen now have to go further afield to make sure that the village survives. A daily reminder that coral and humanity are intrinsically and forever linked.

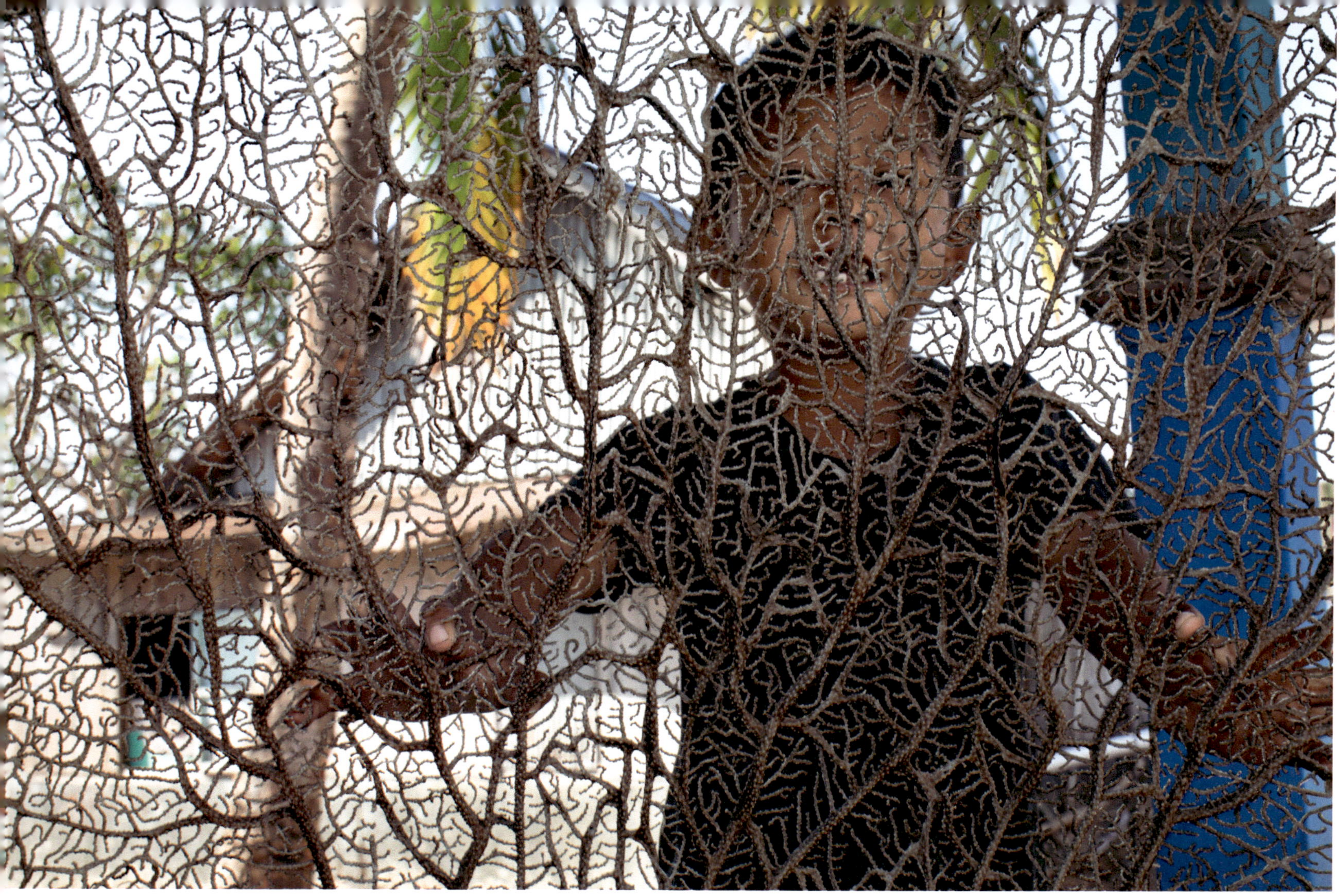

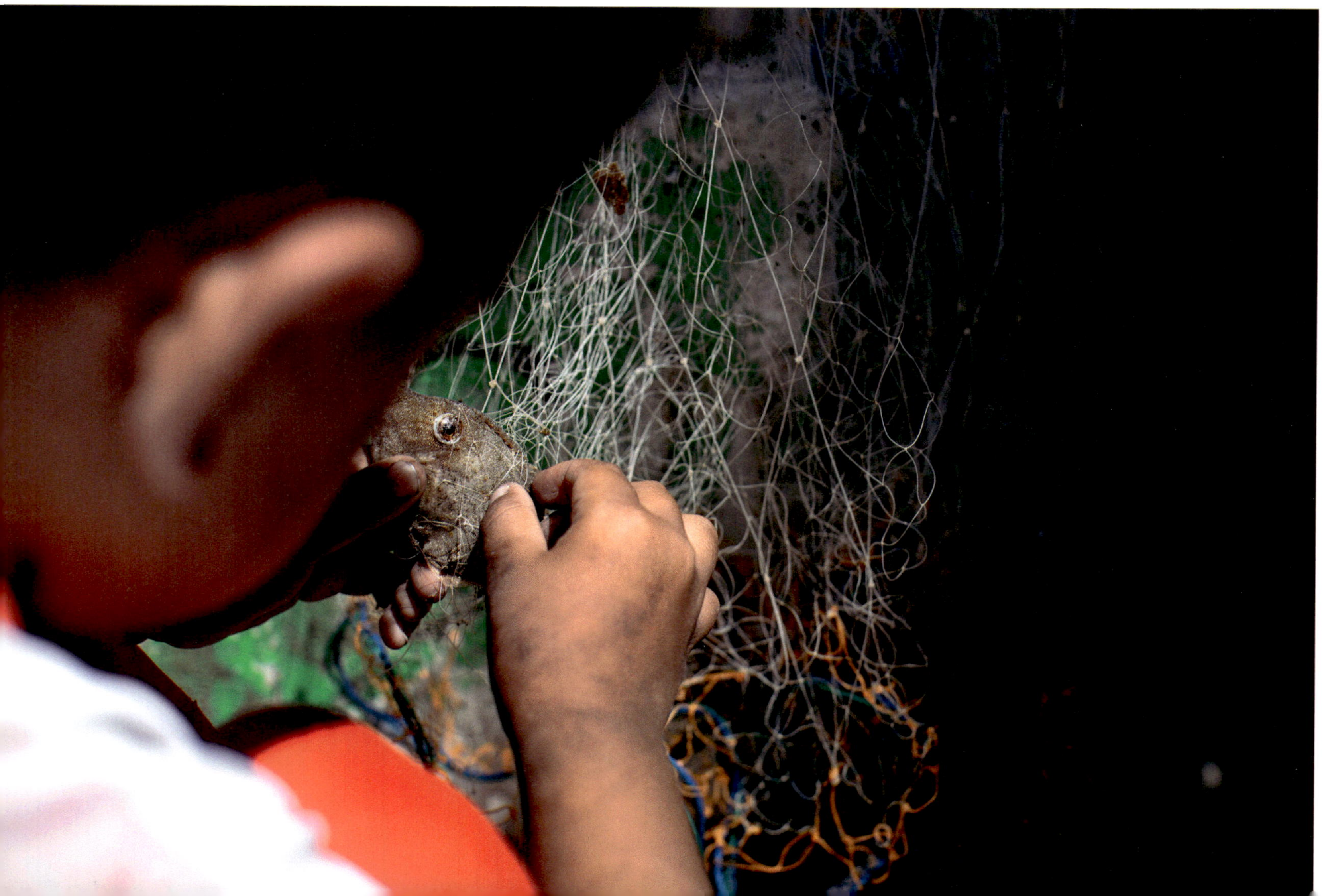

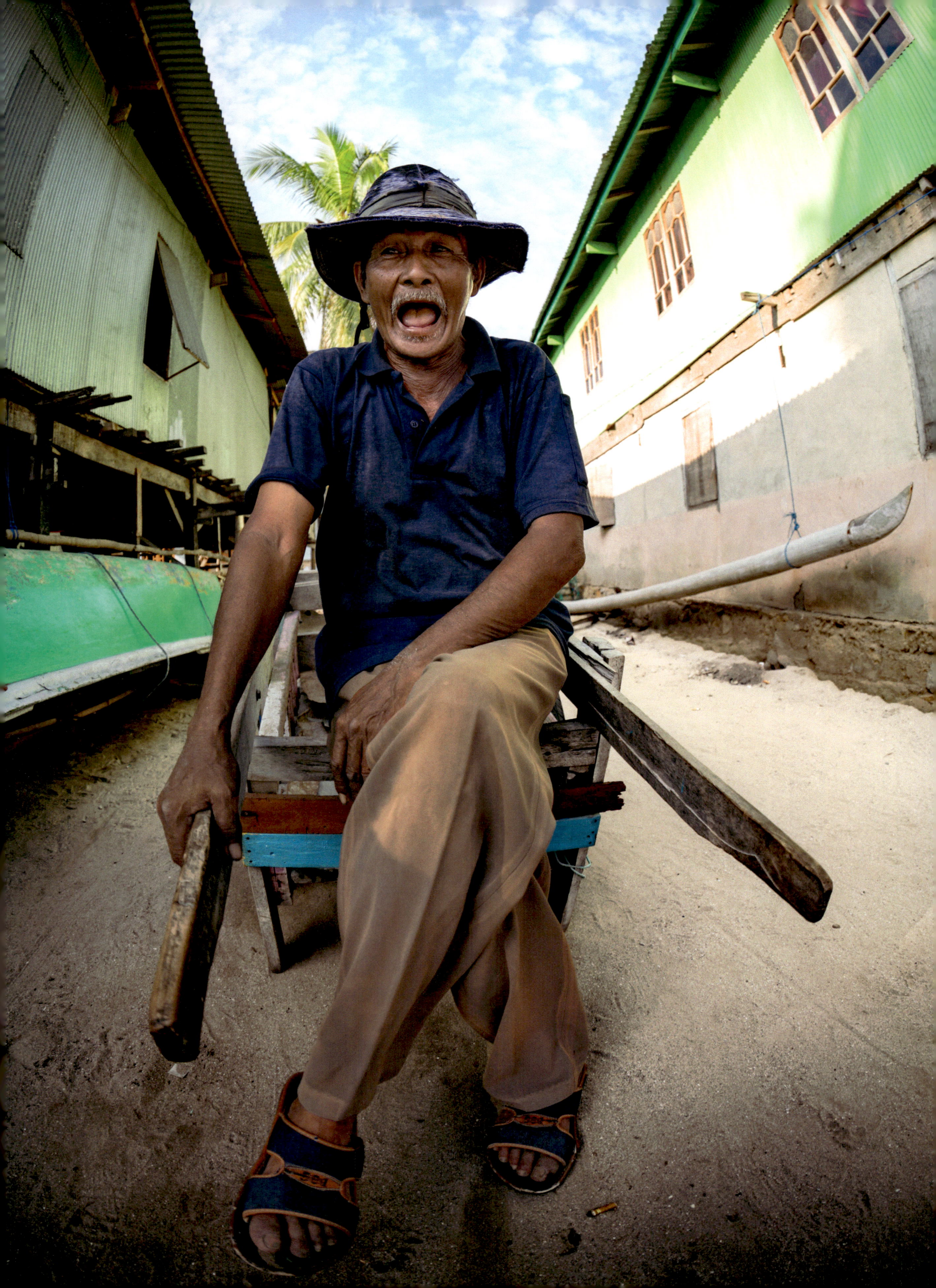

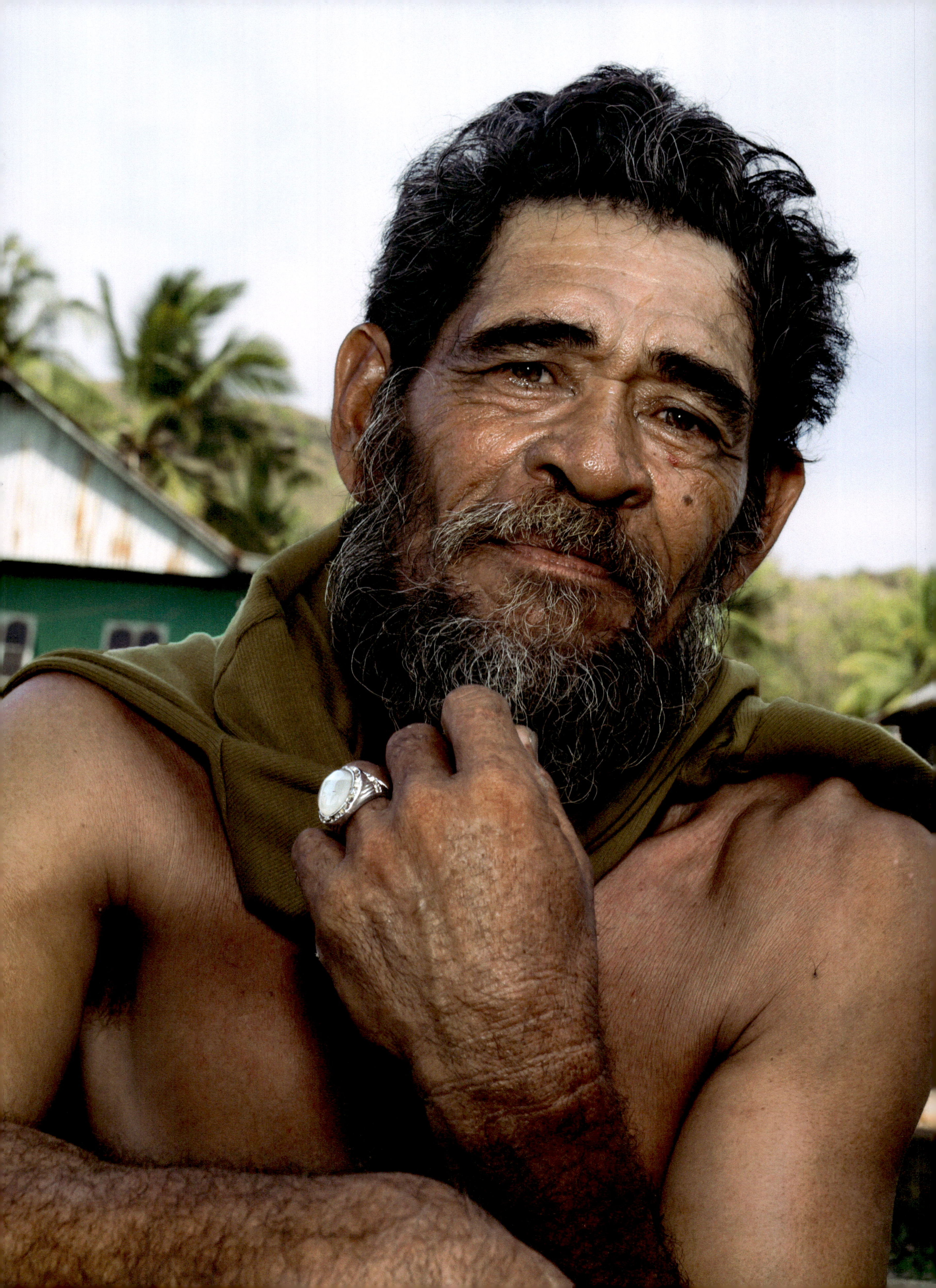

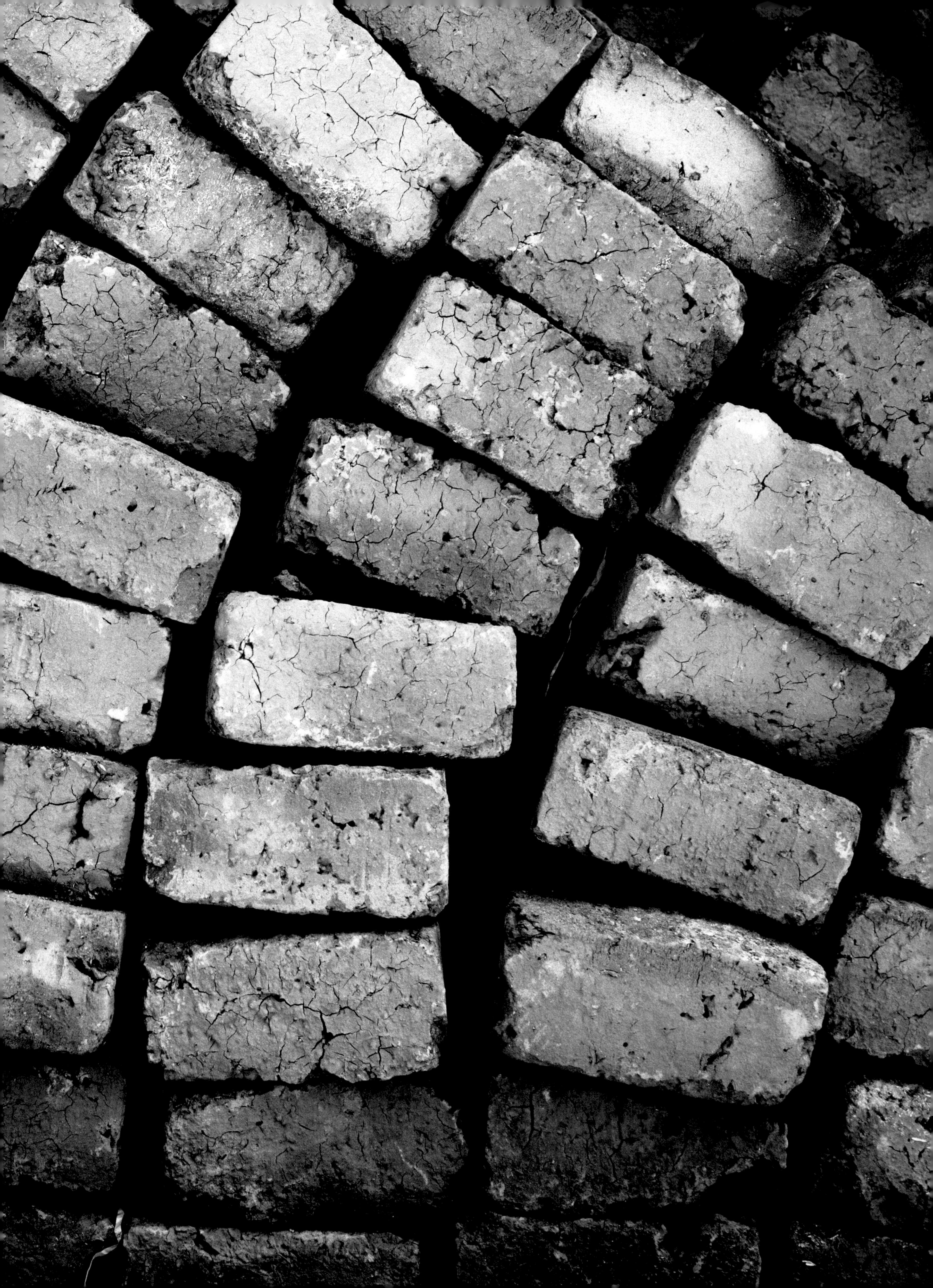

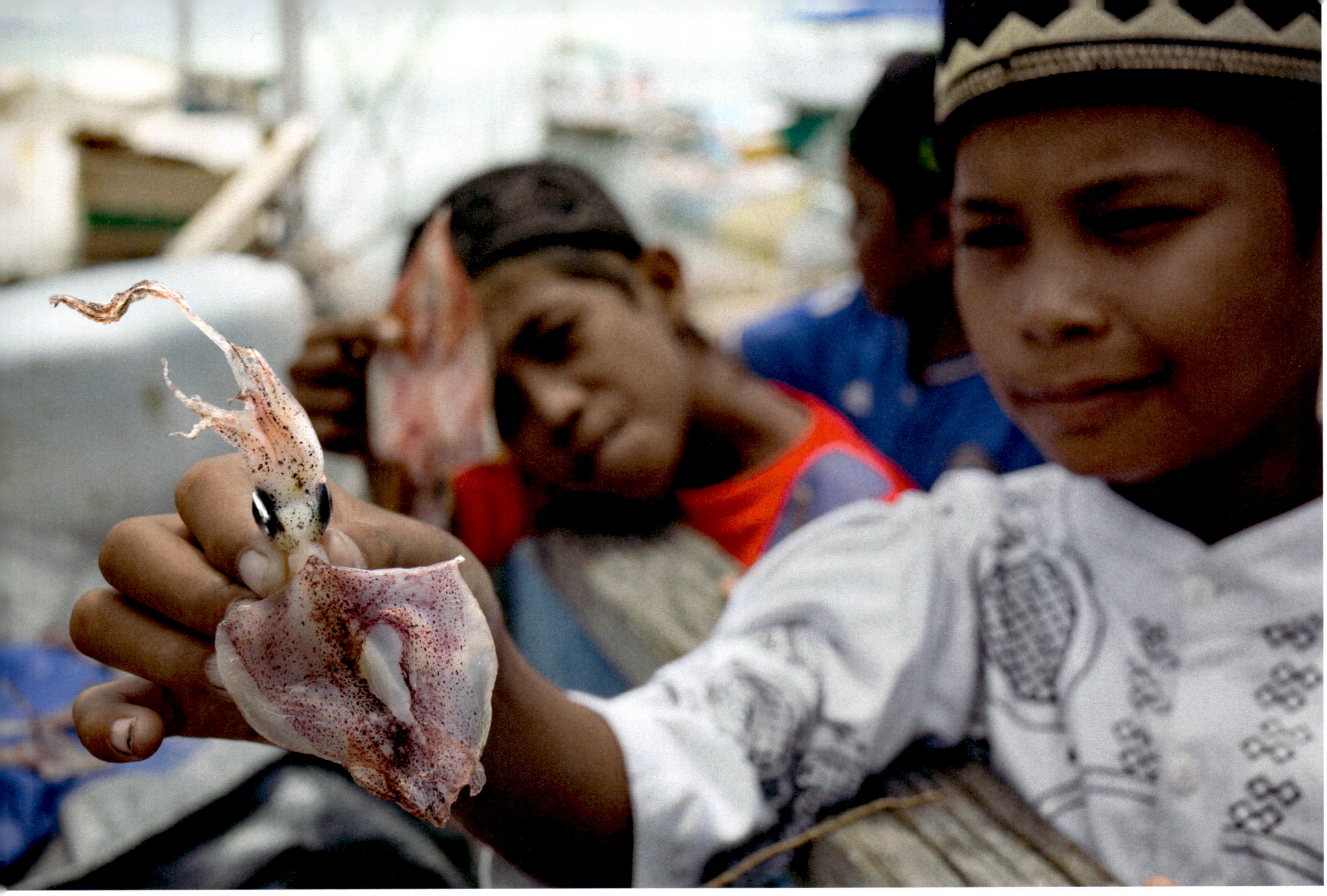

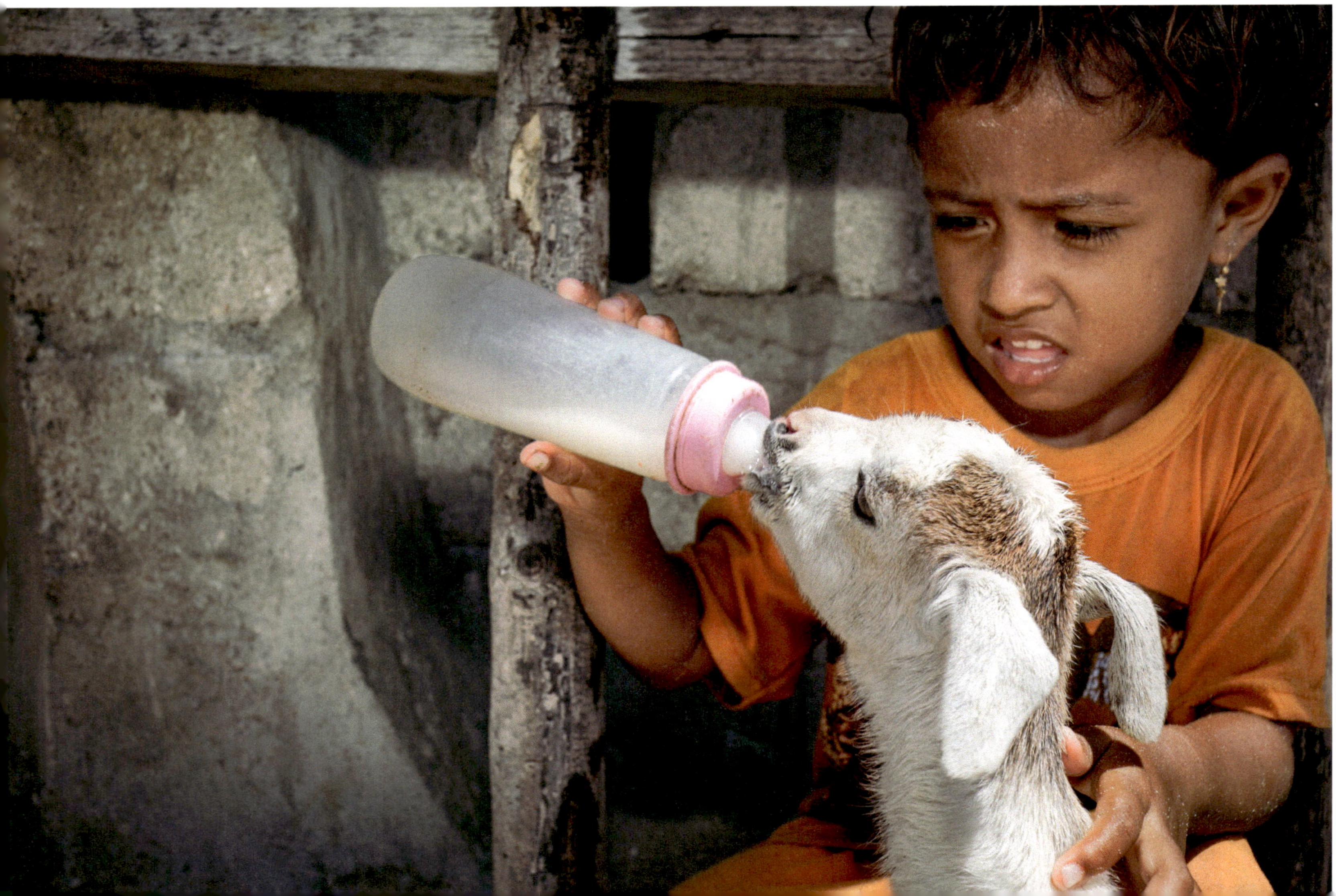

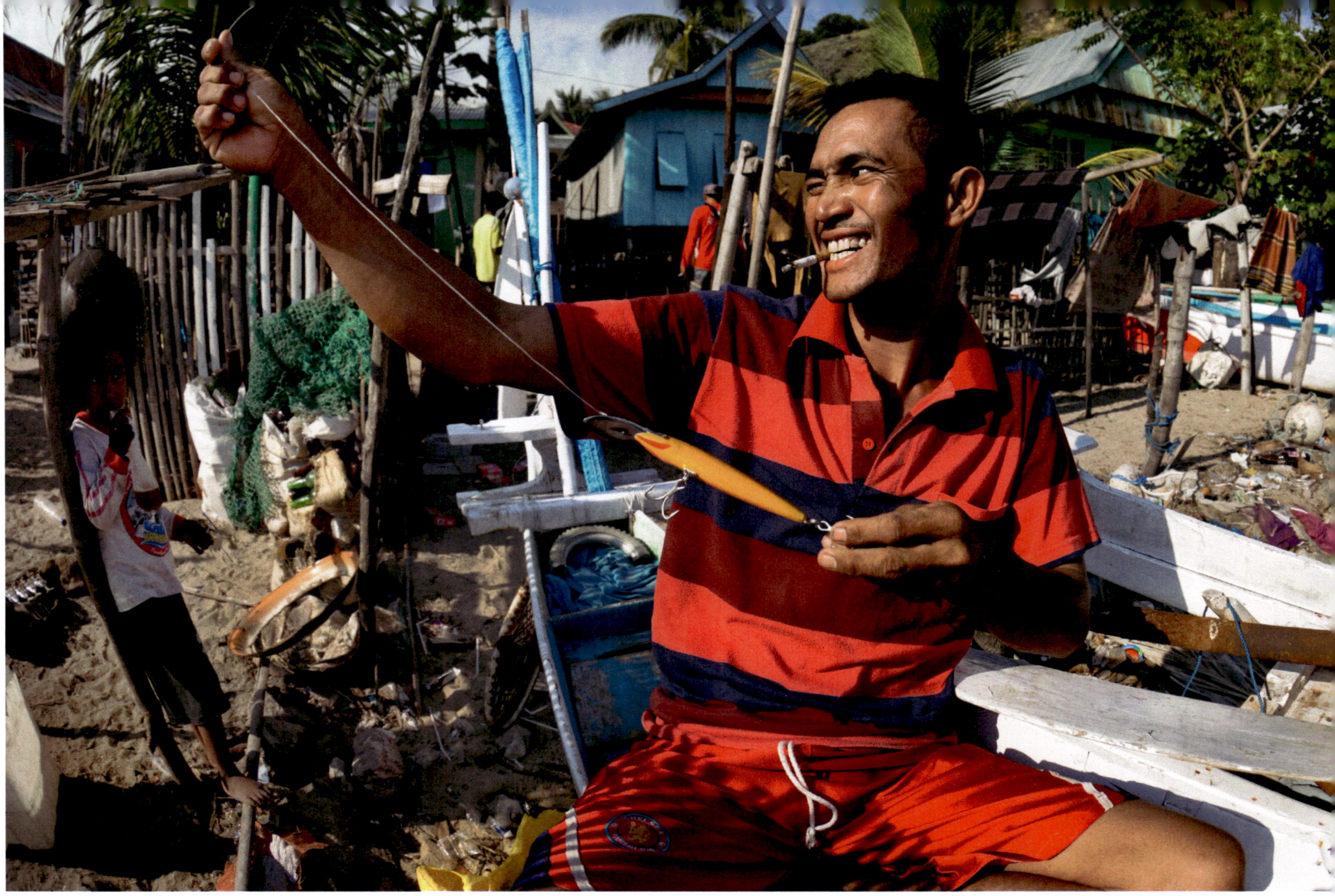

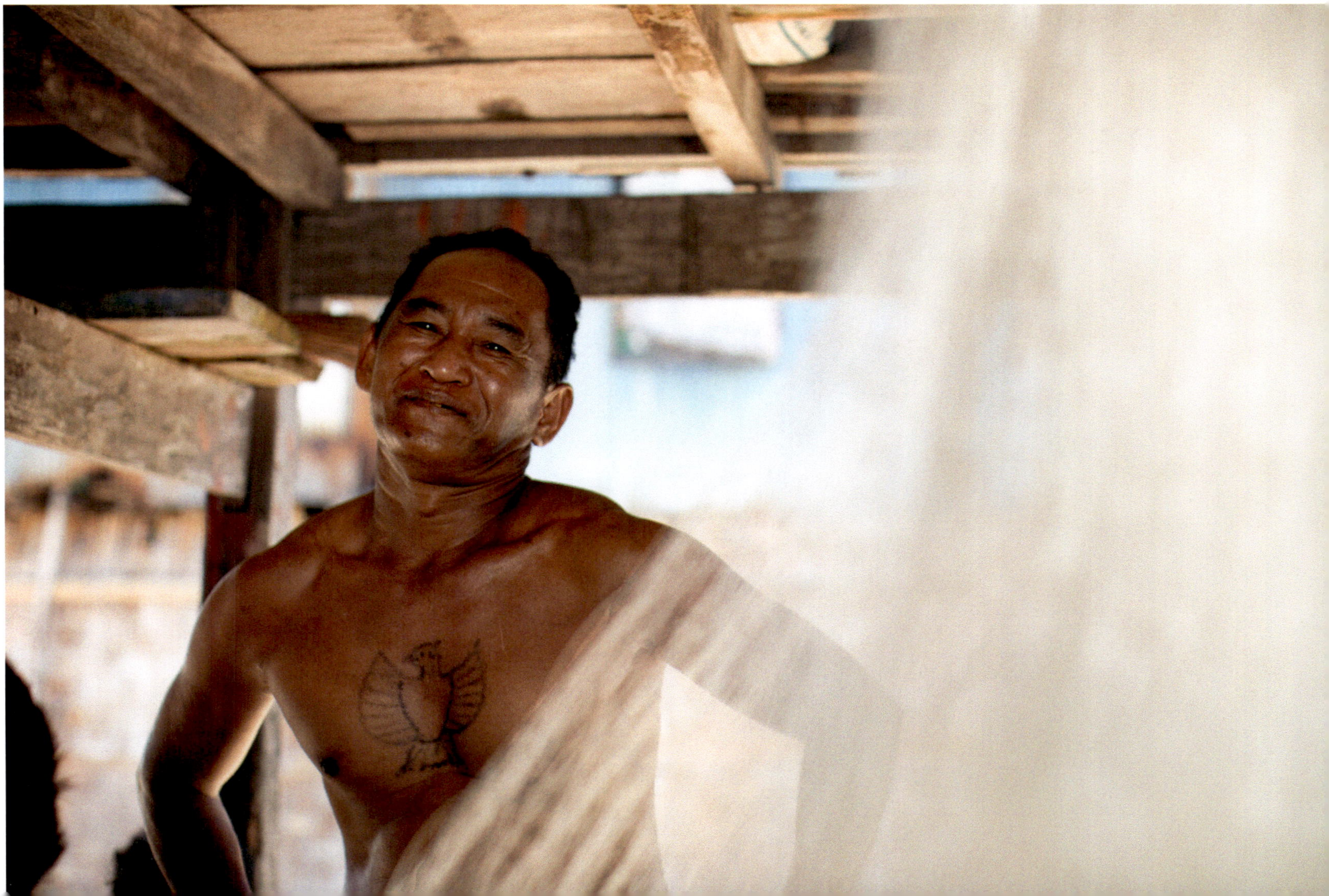

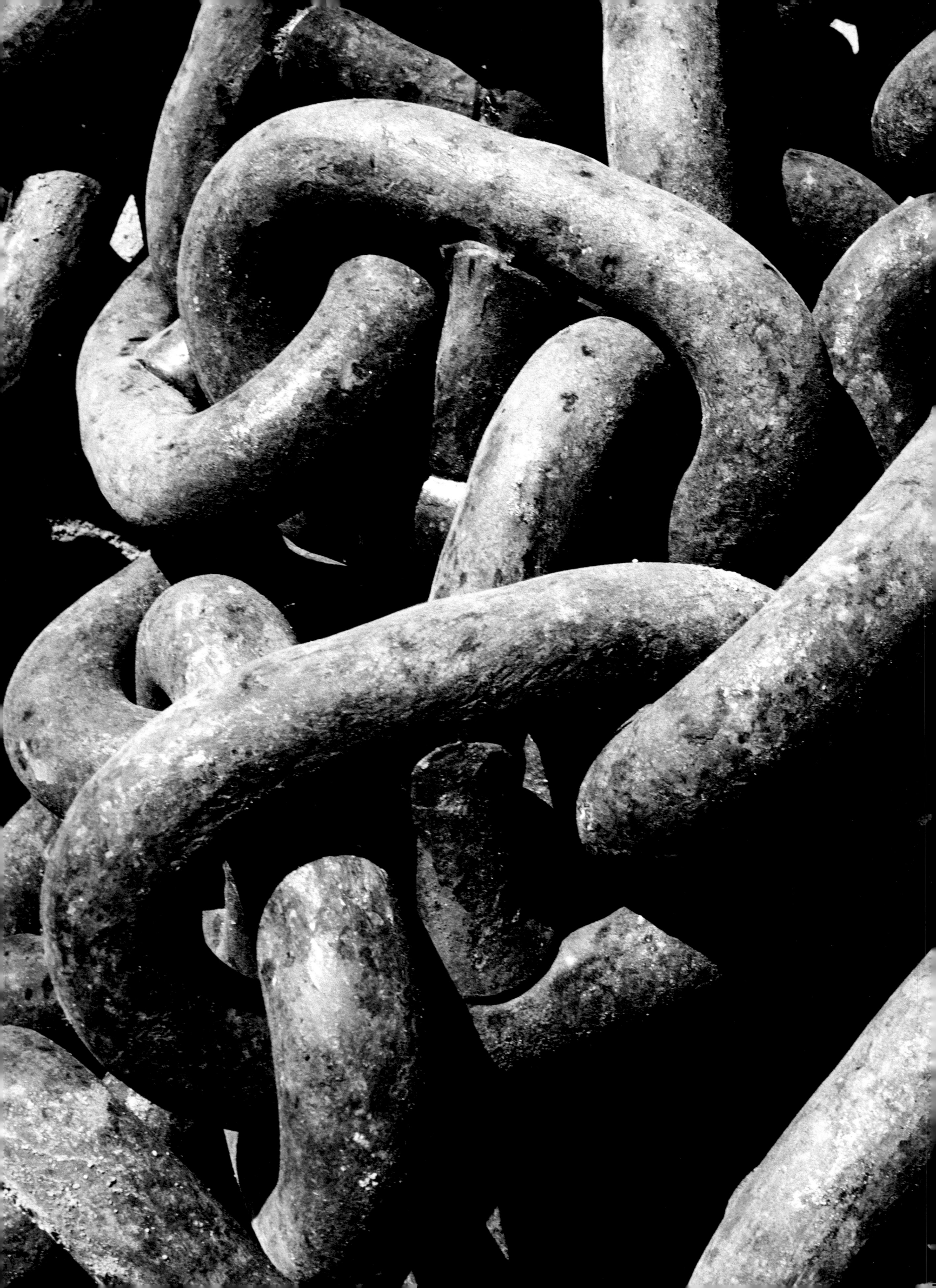

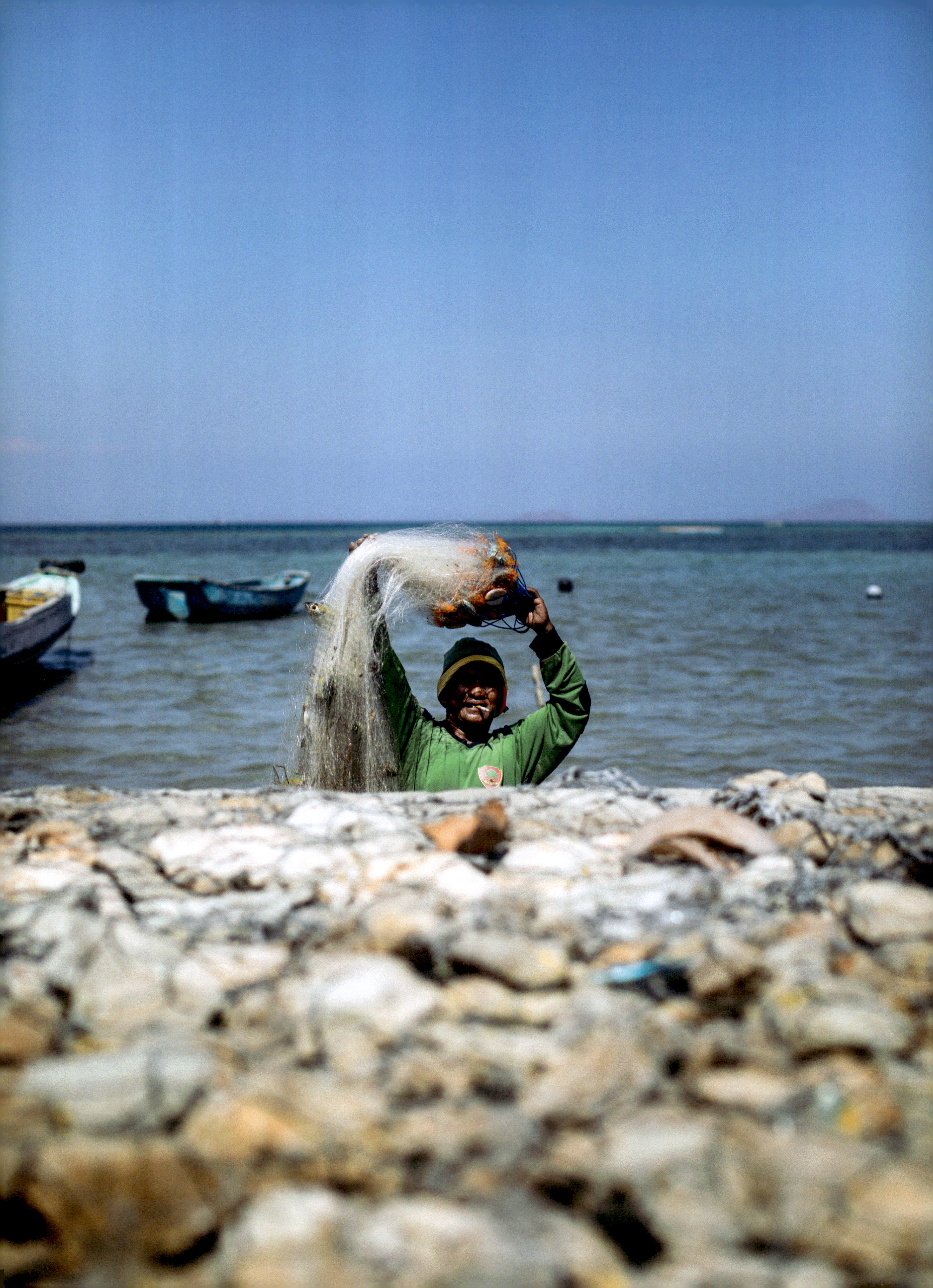

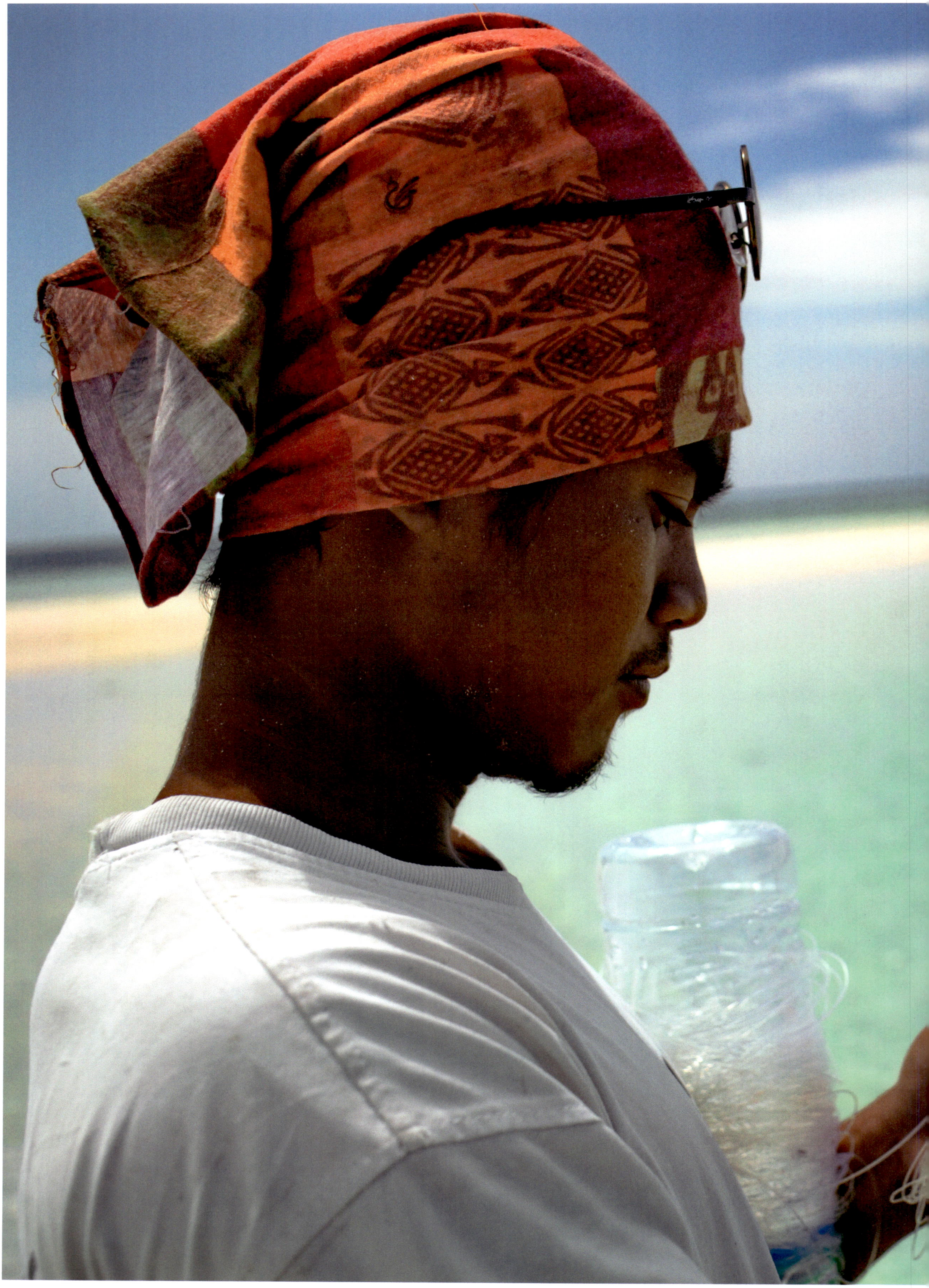

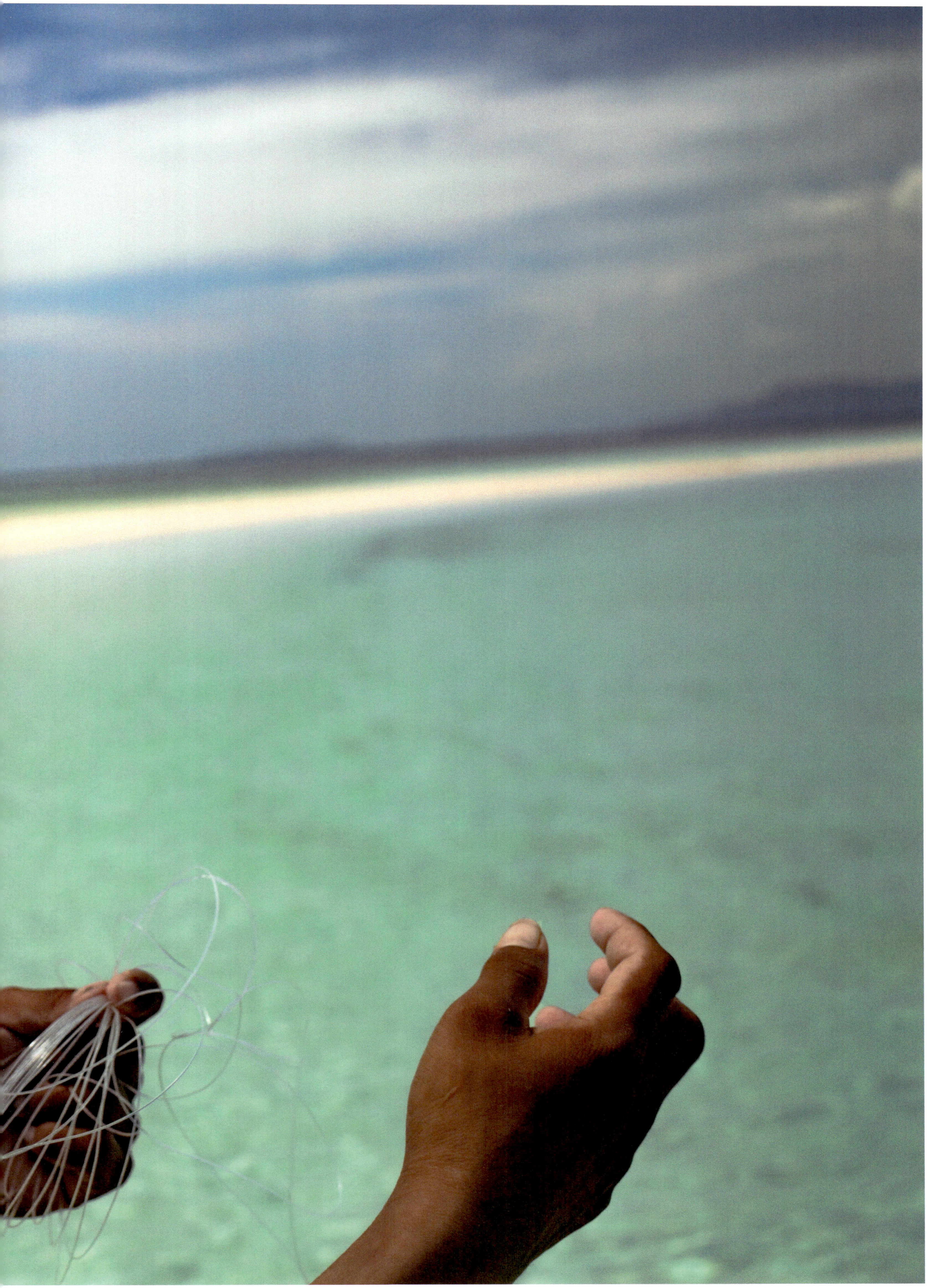

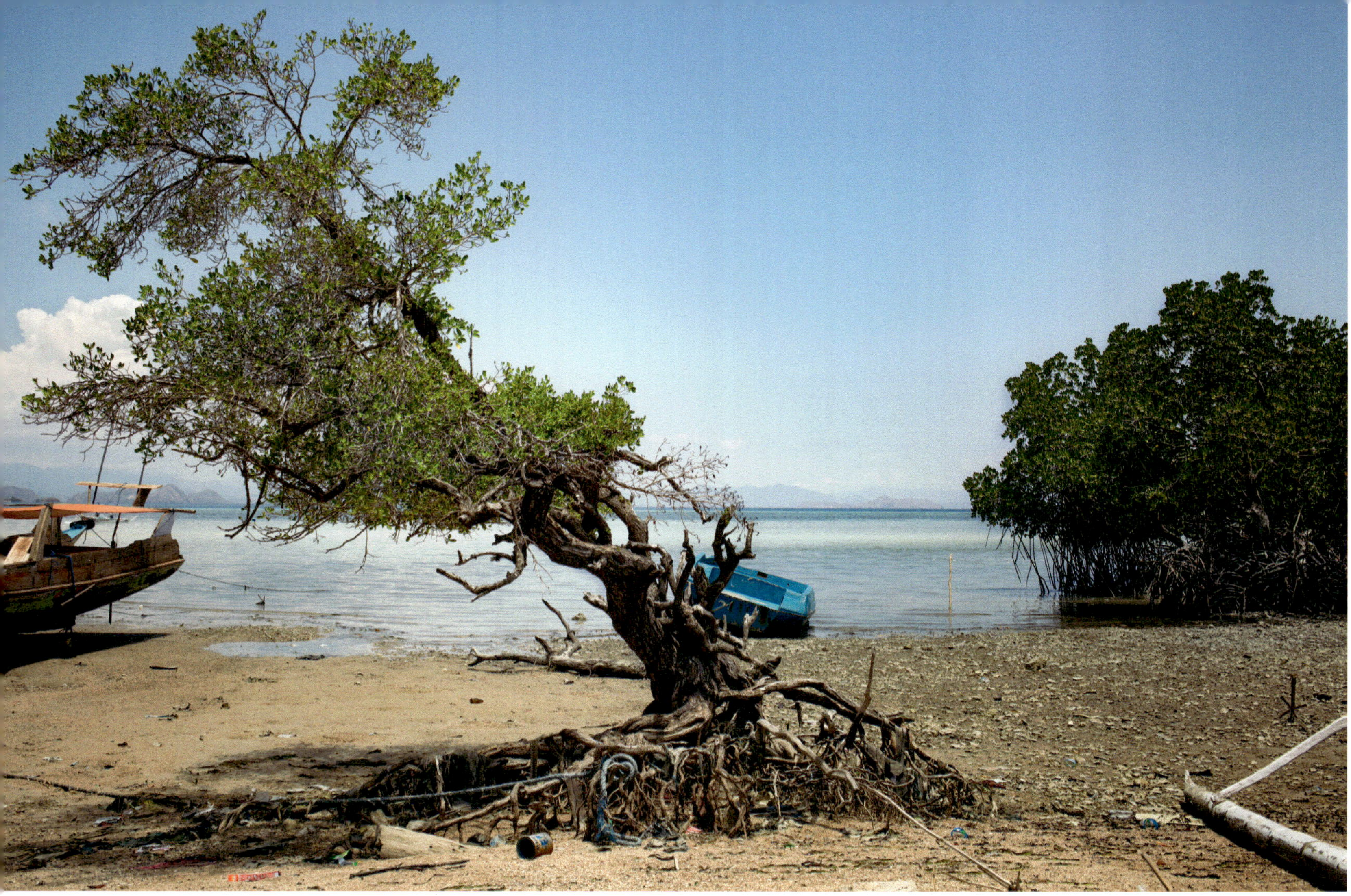

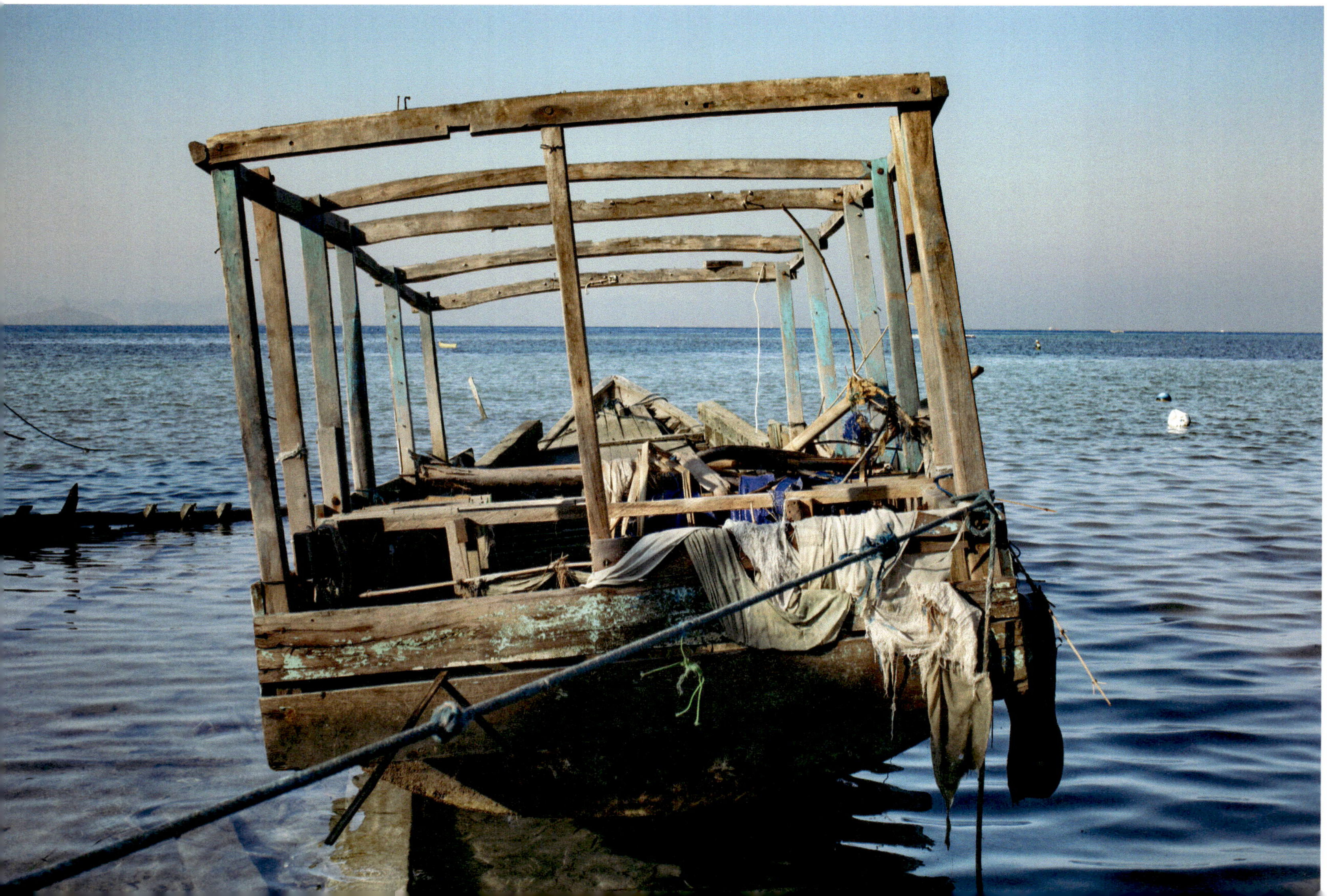

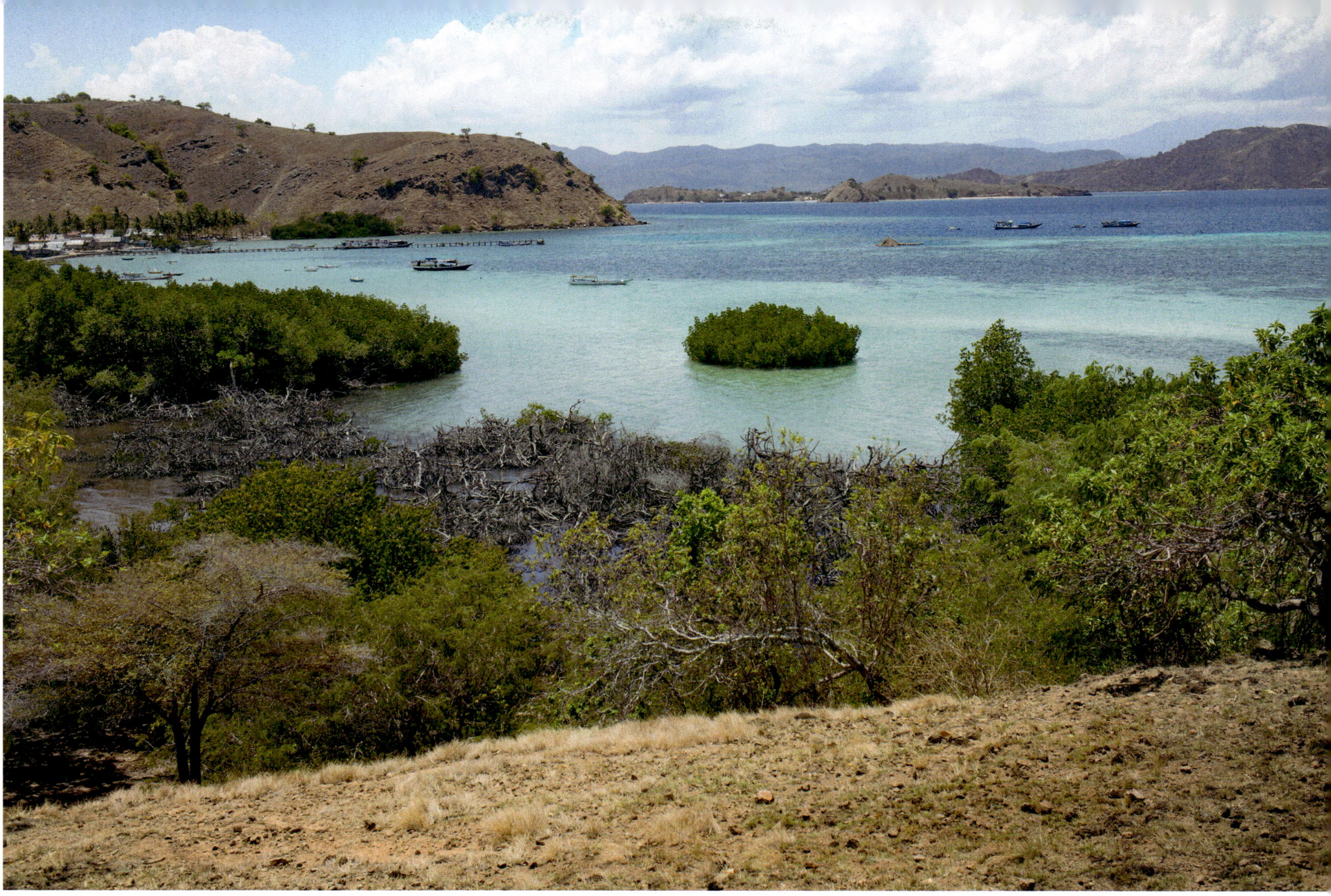

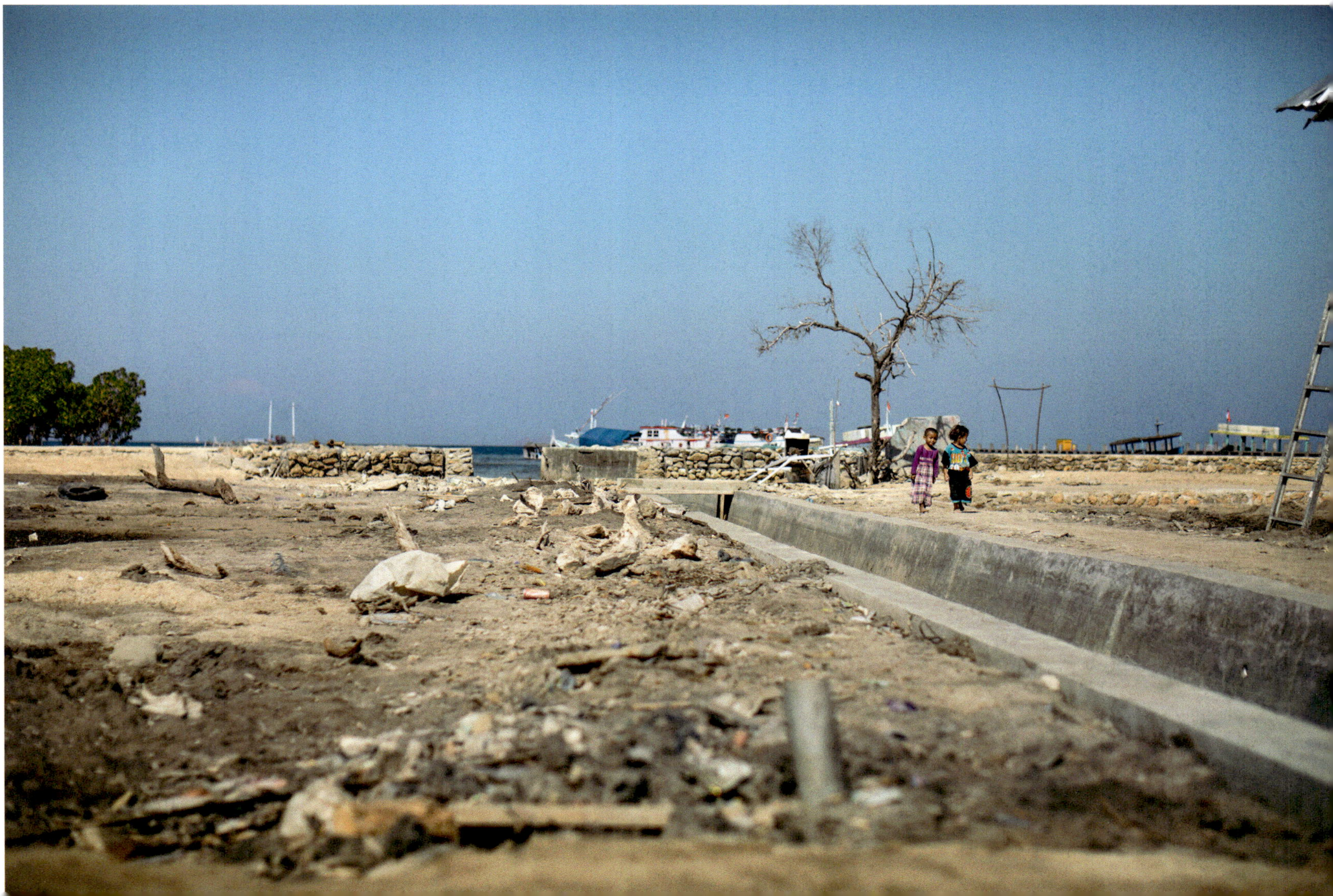

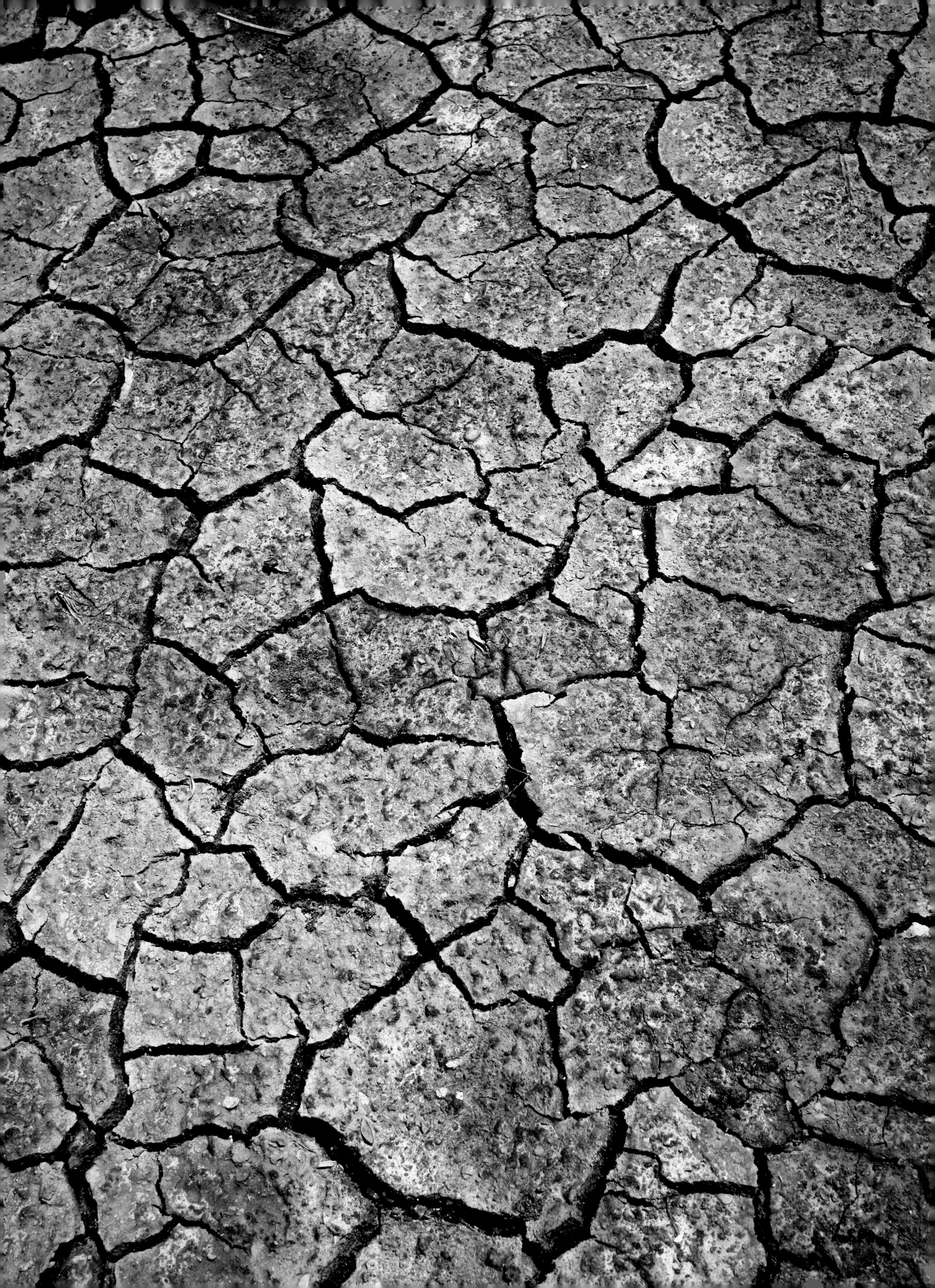

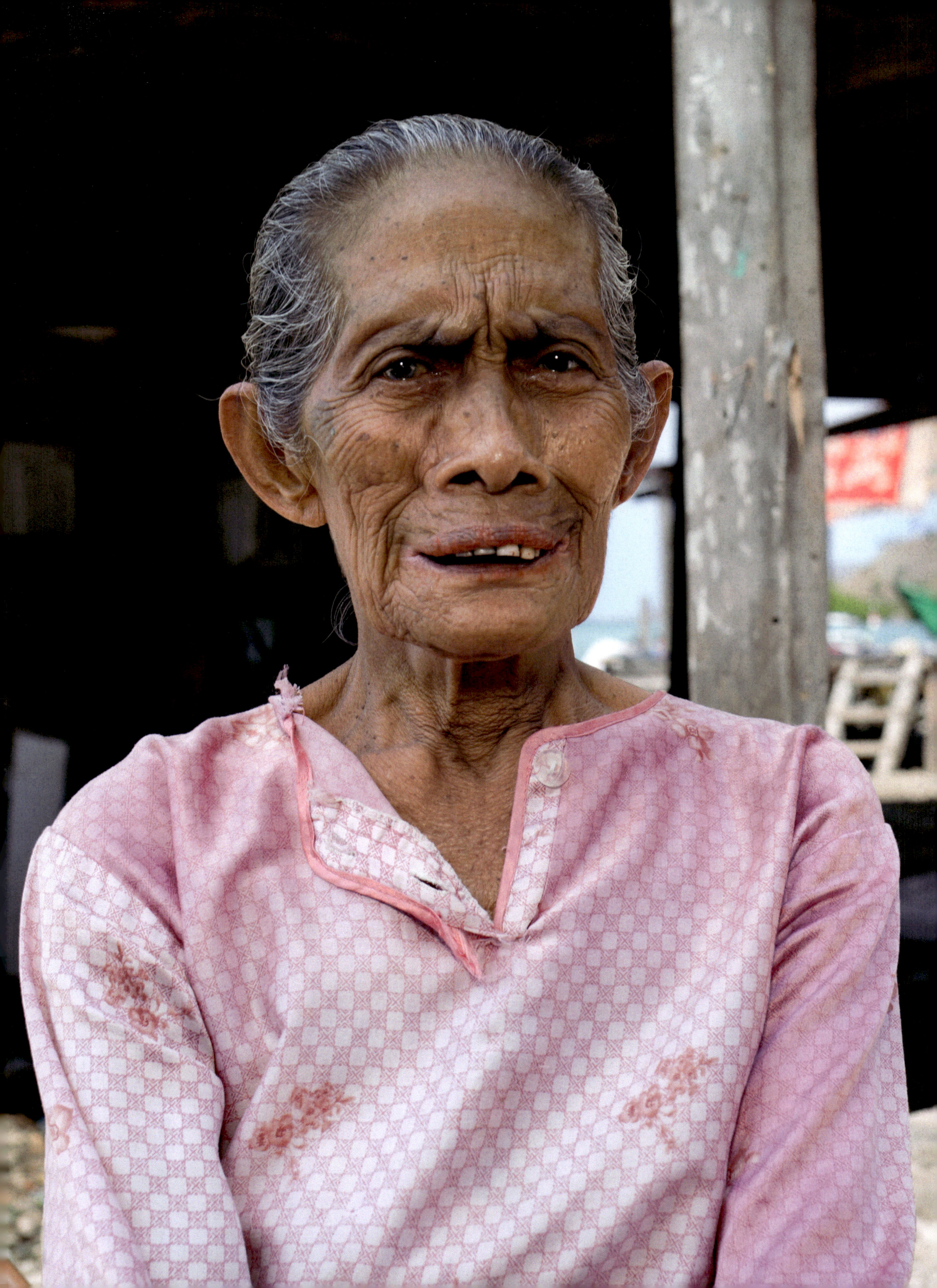

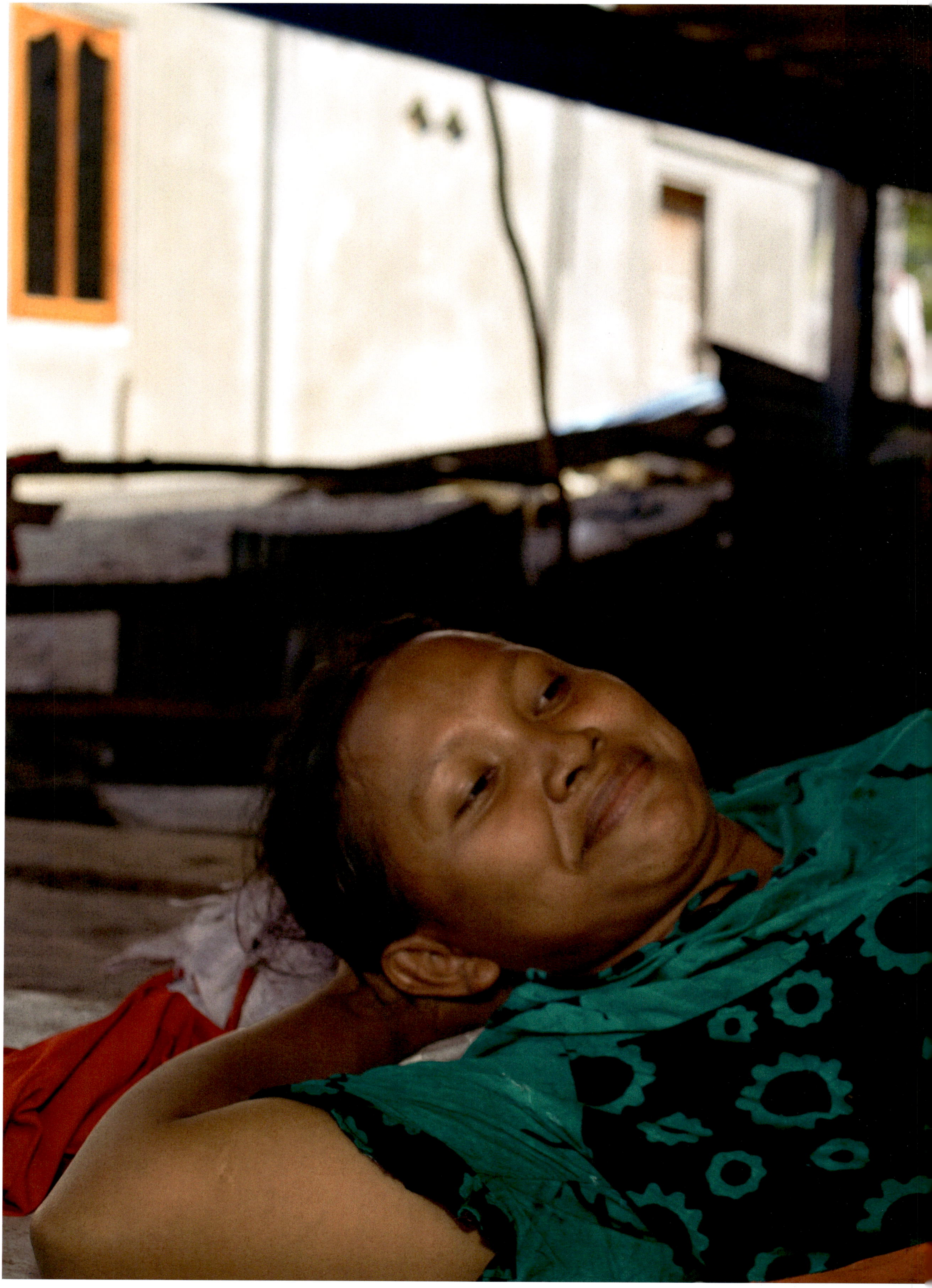

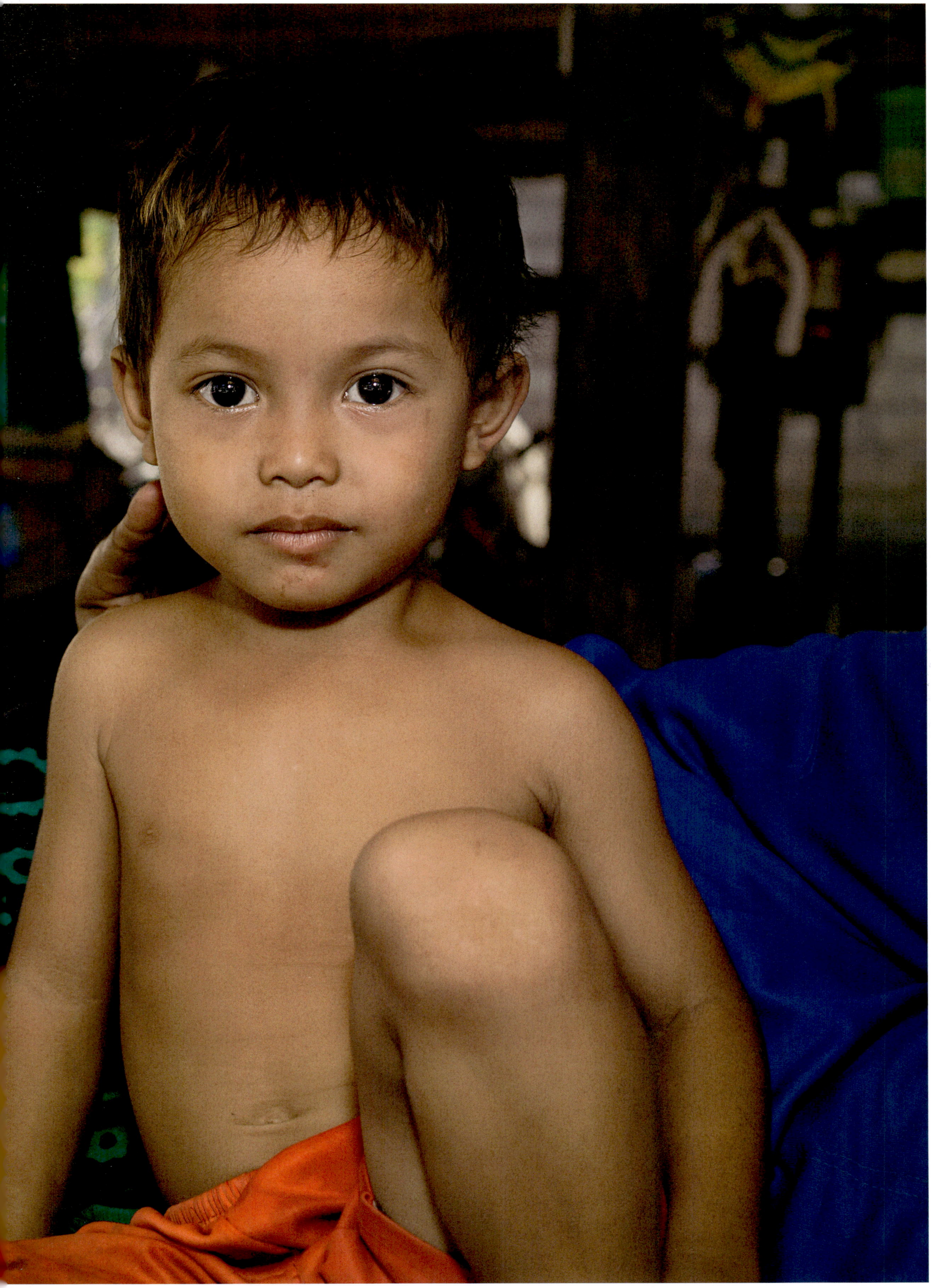

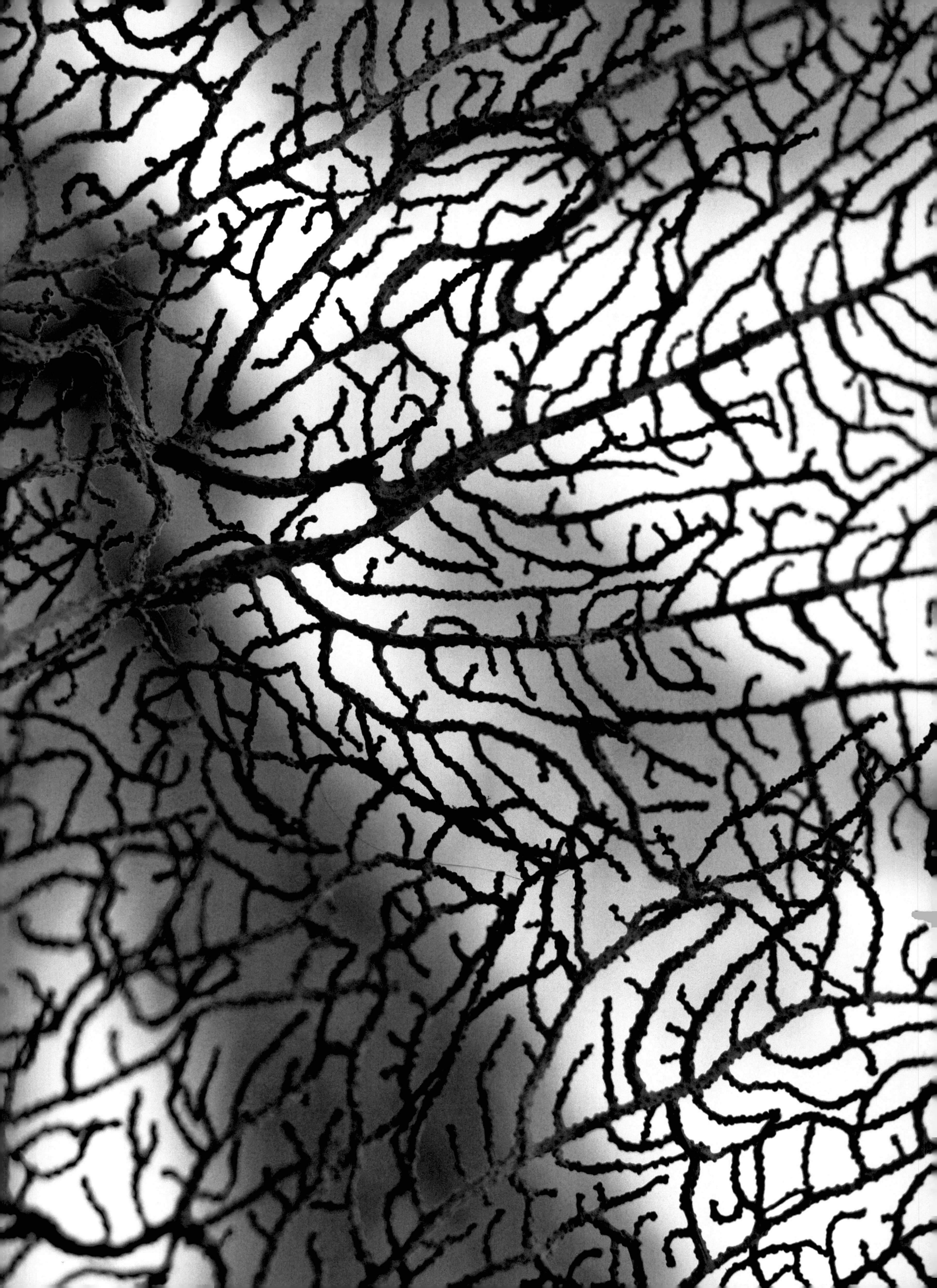

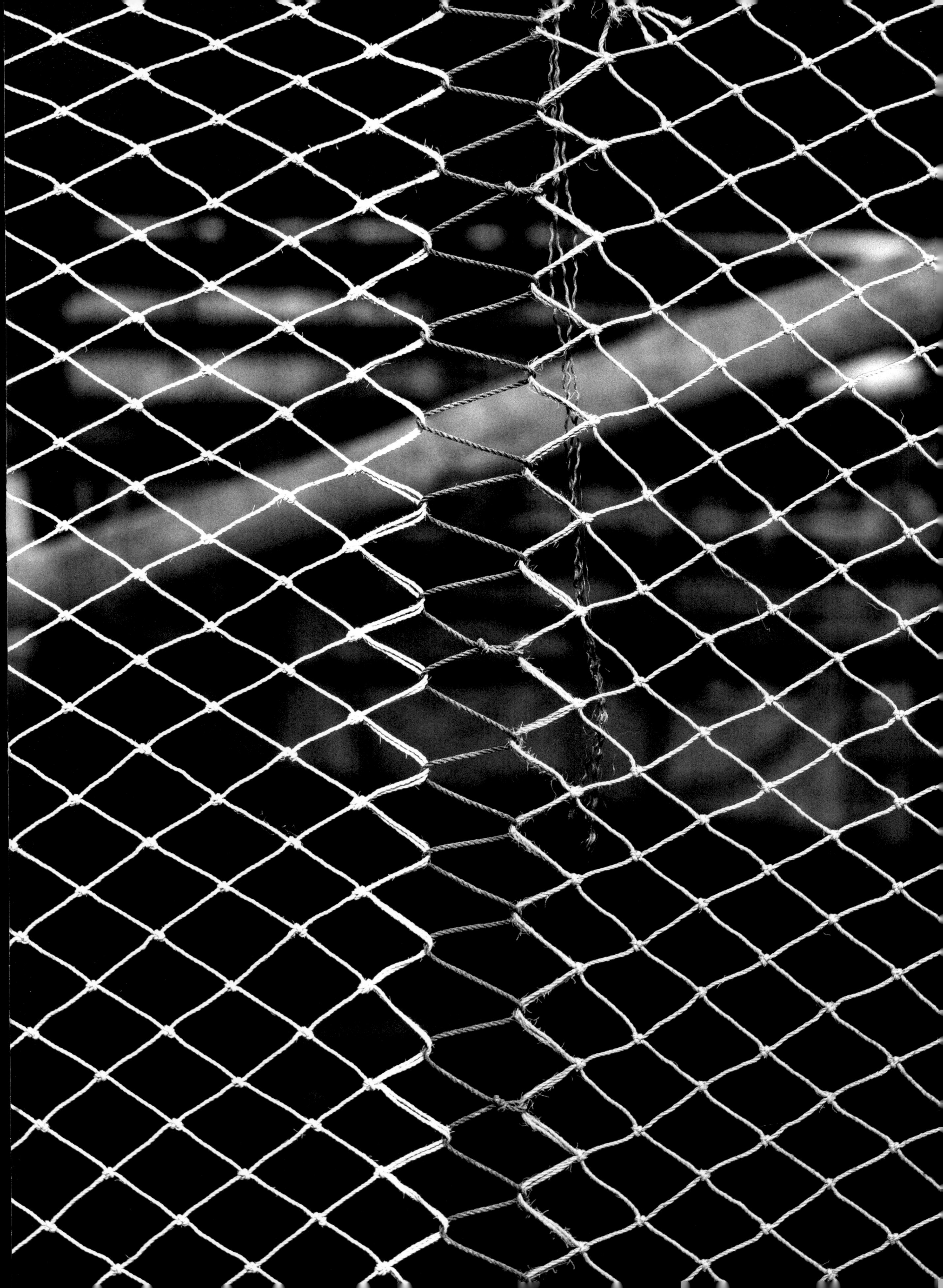

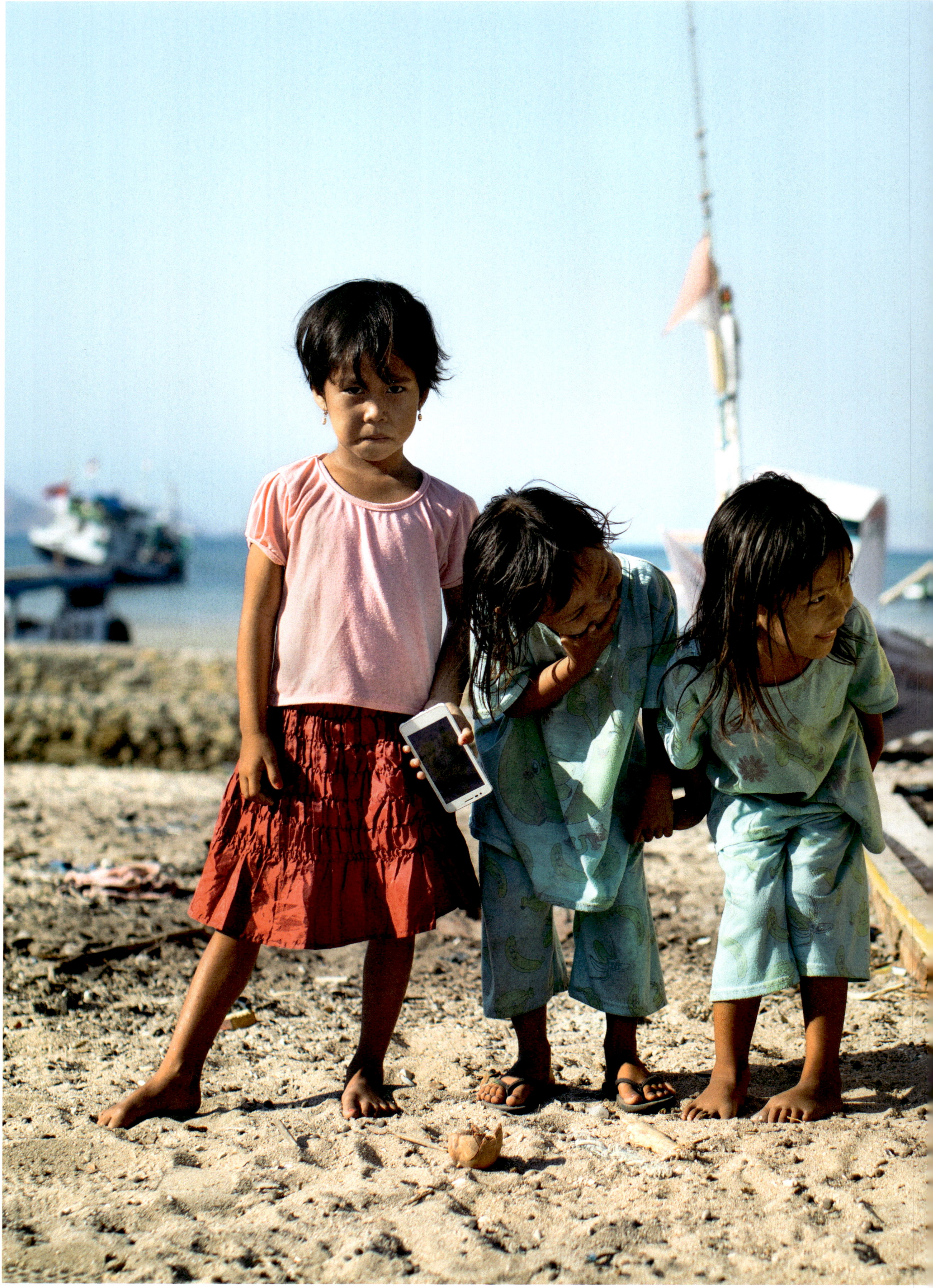

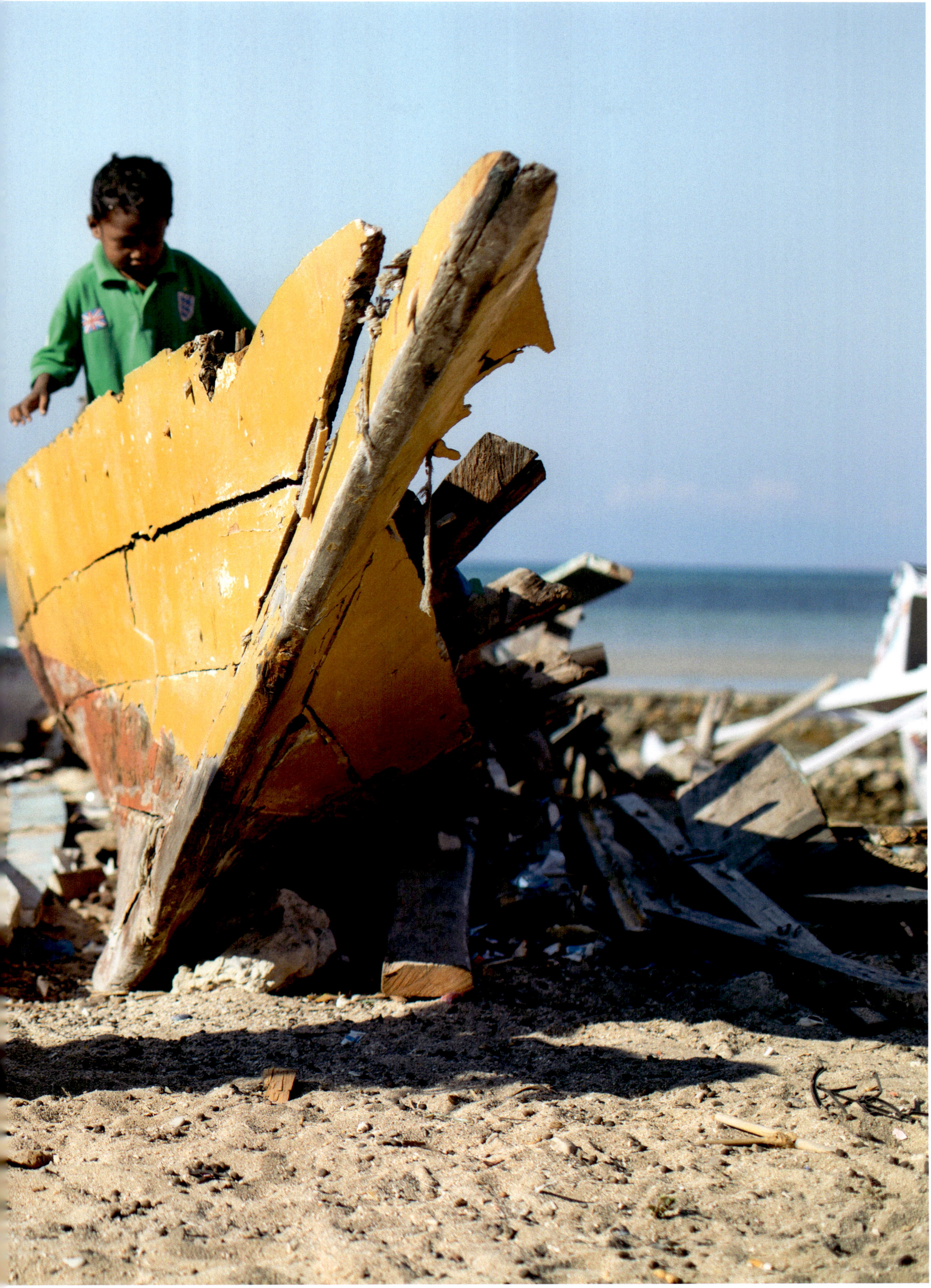

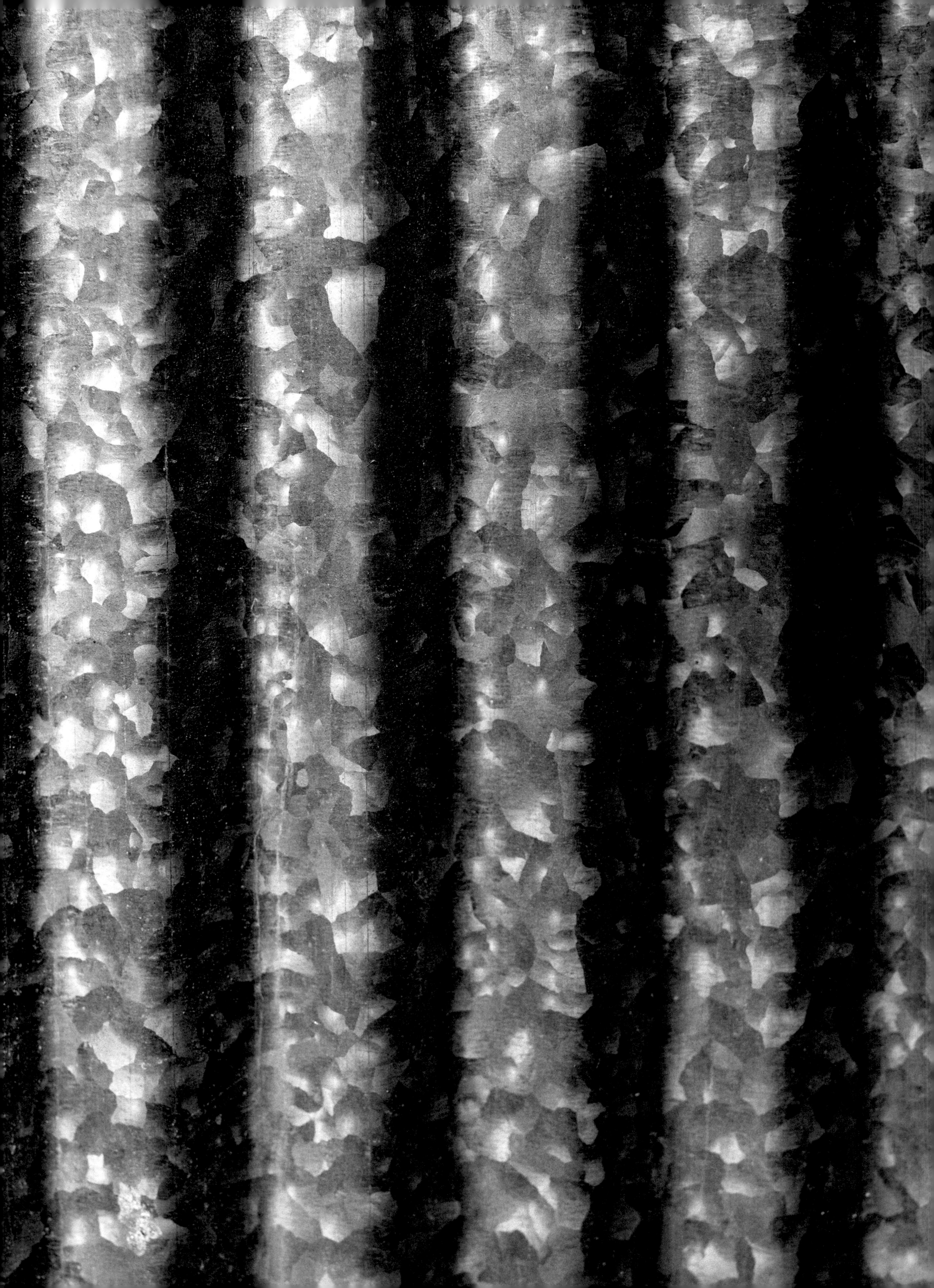

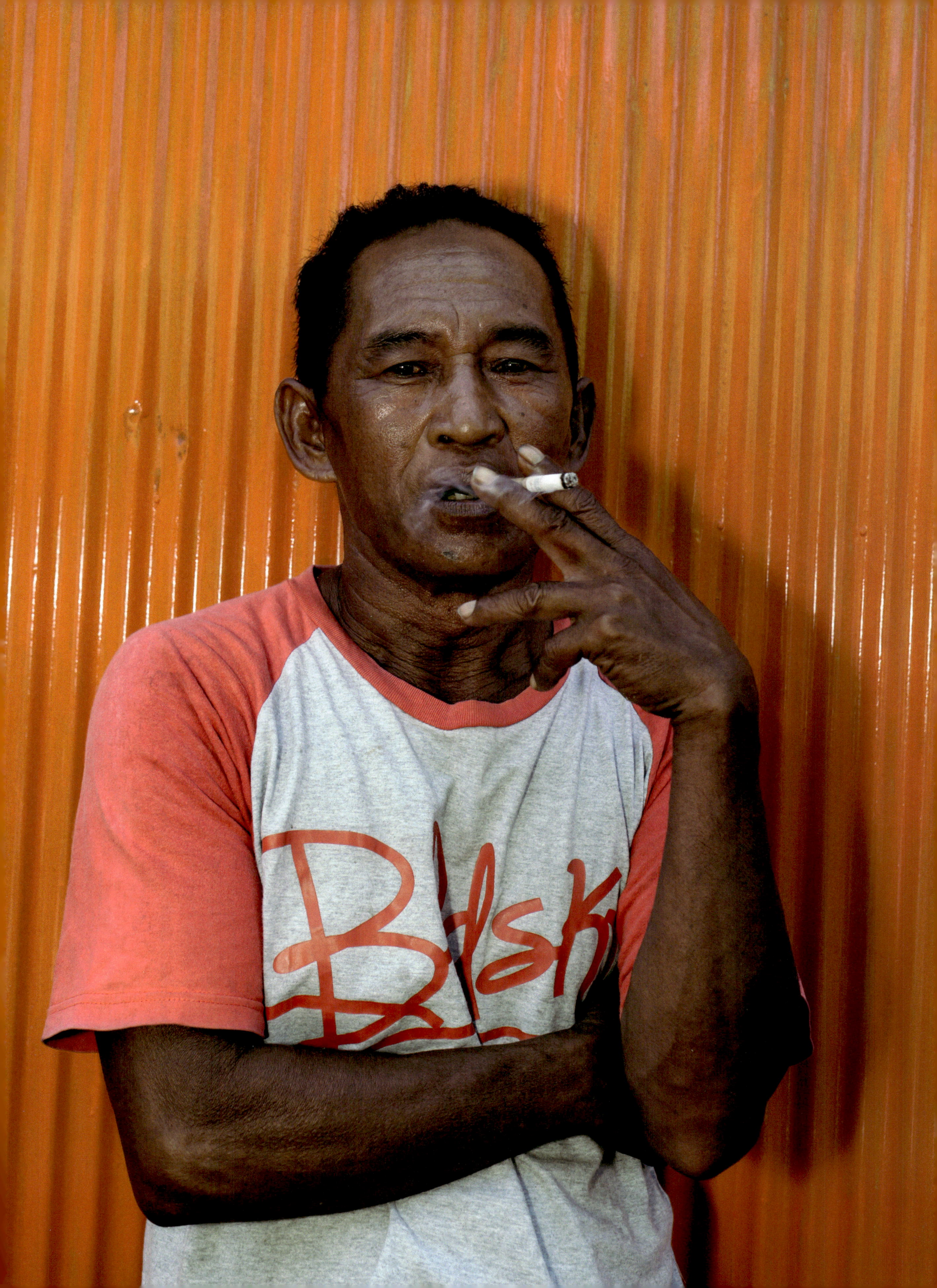

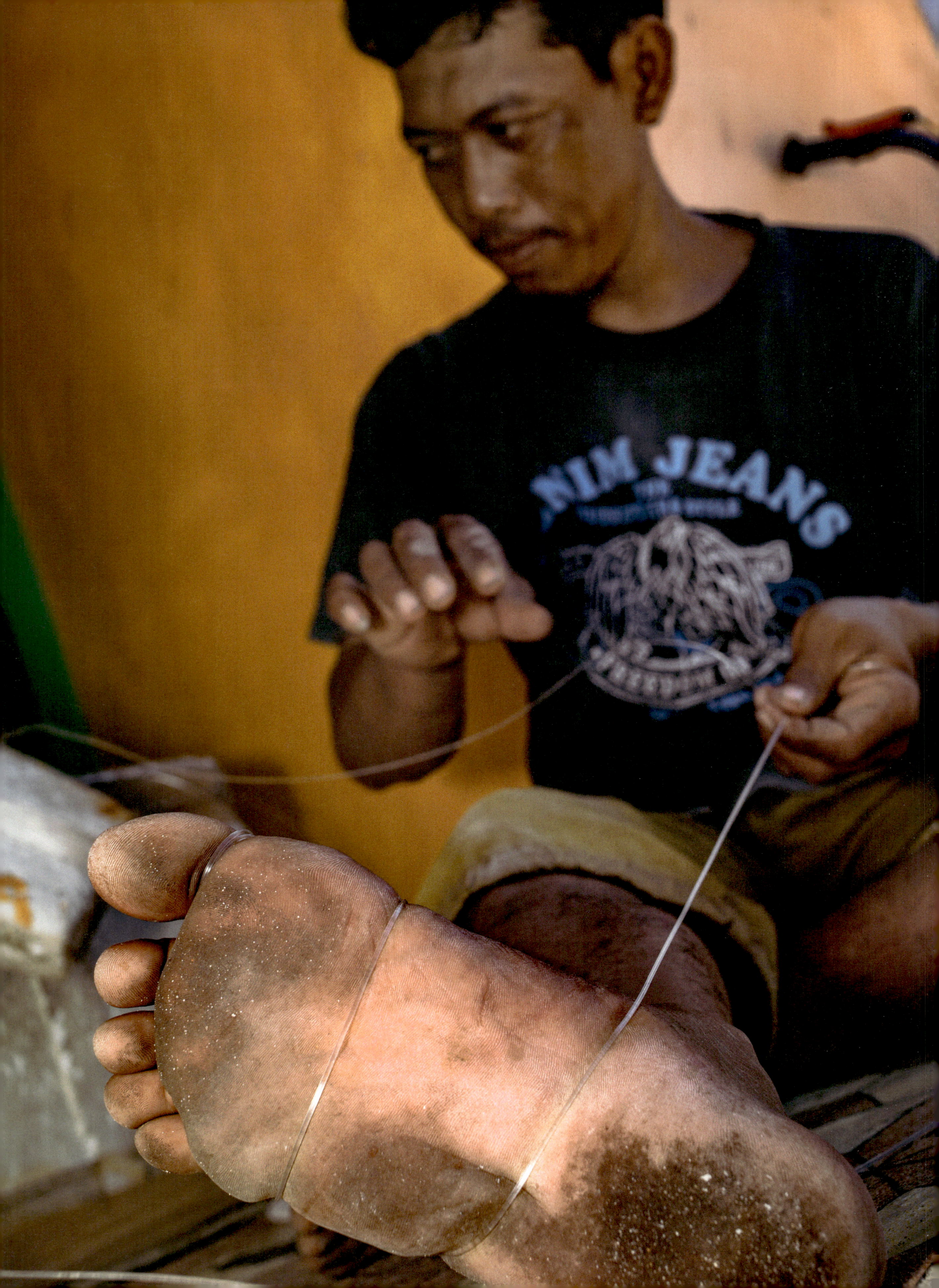
NIM JEANS

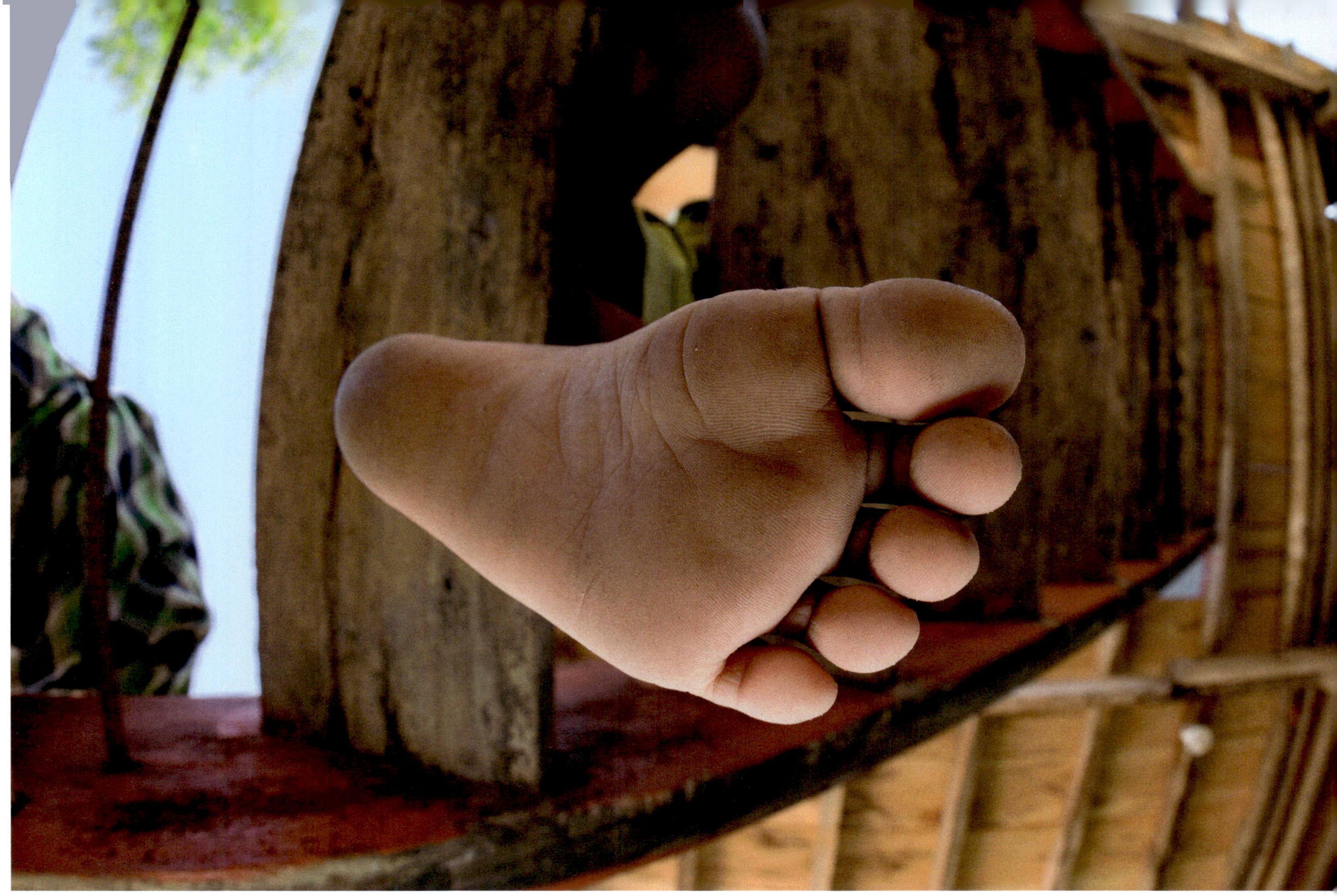

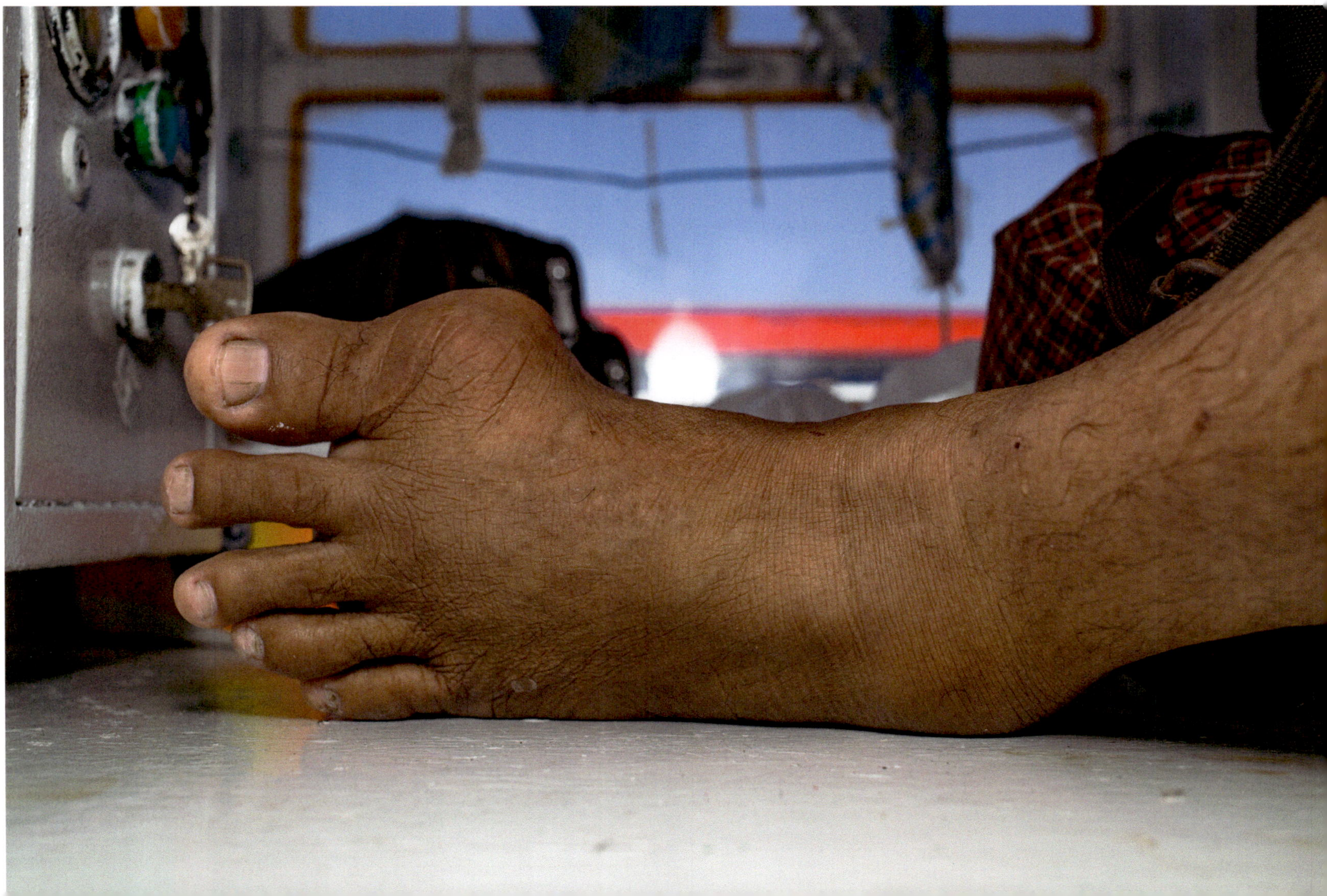

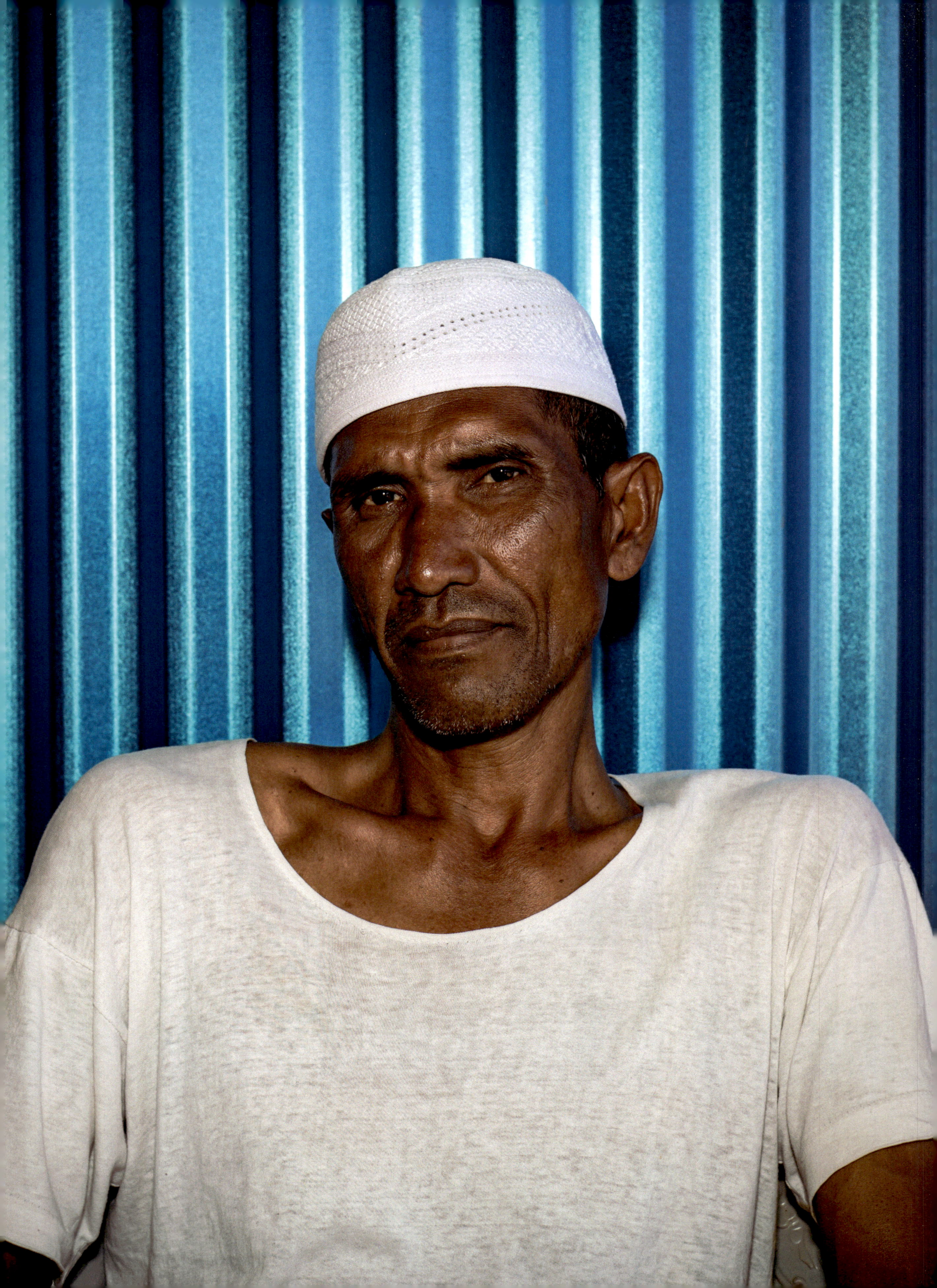

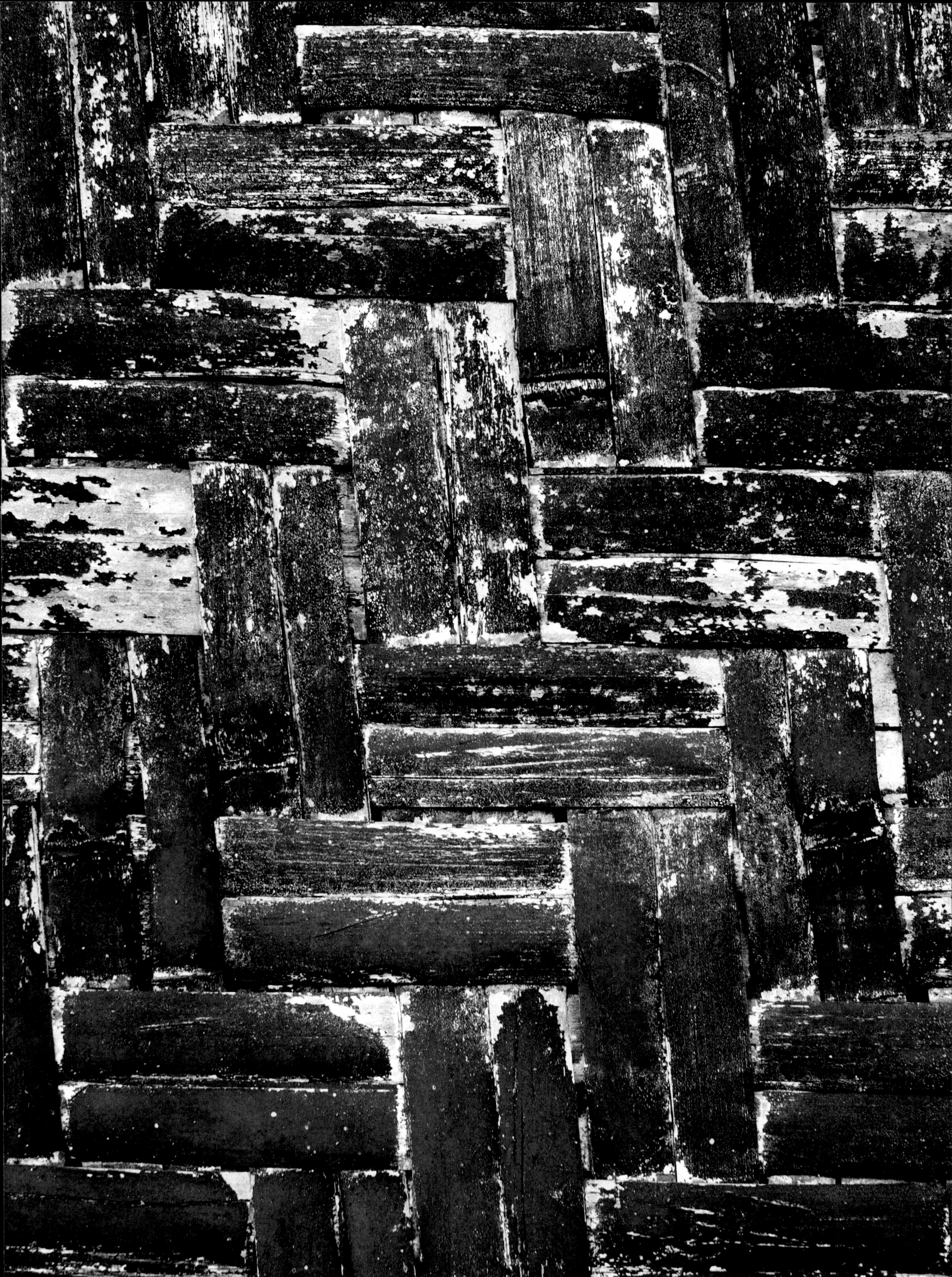

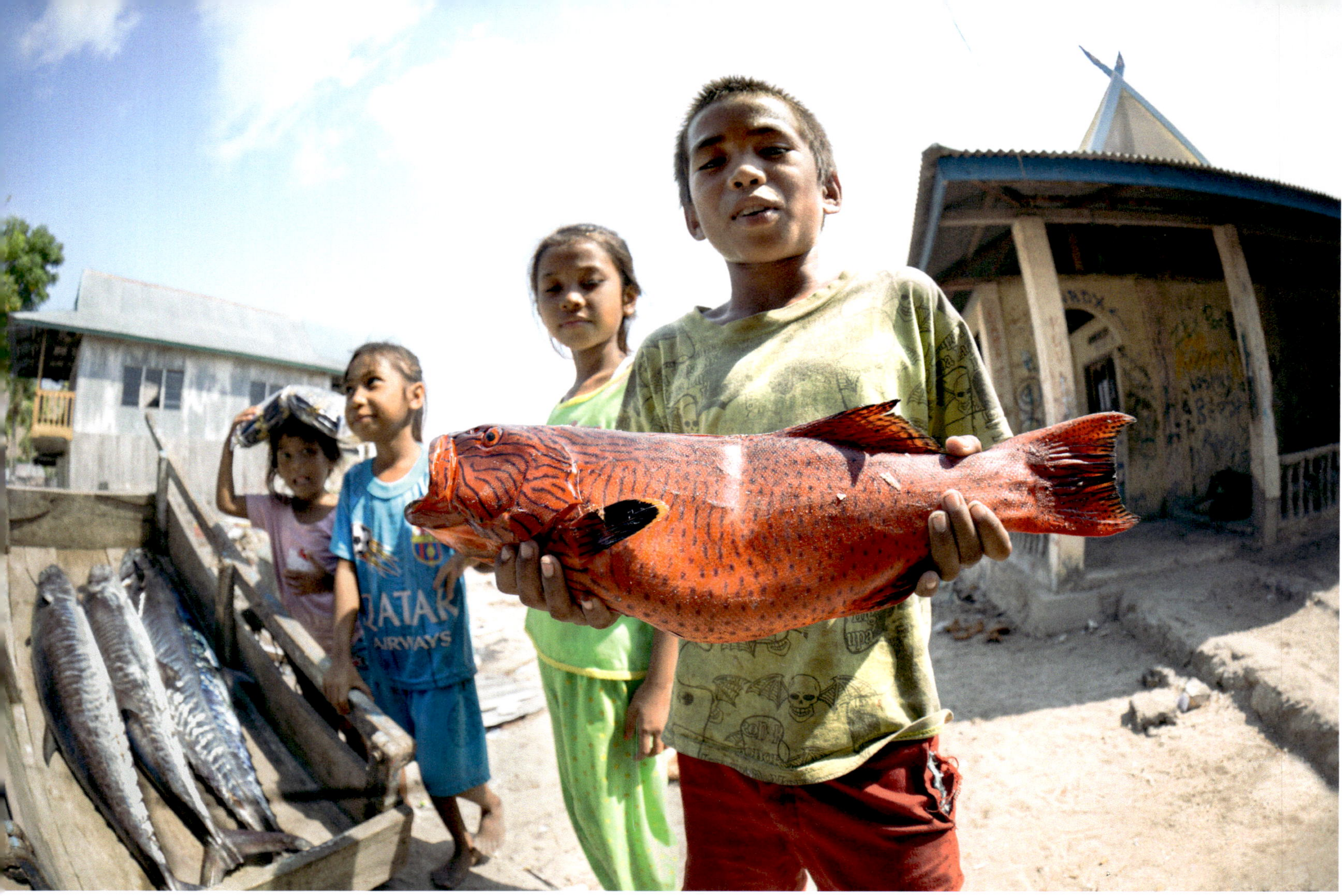

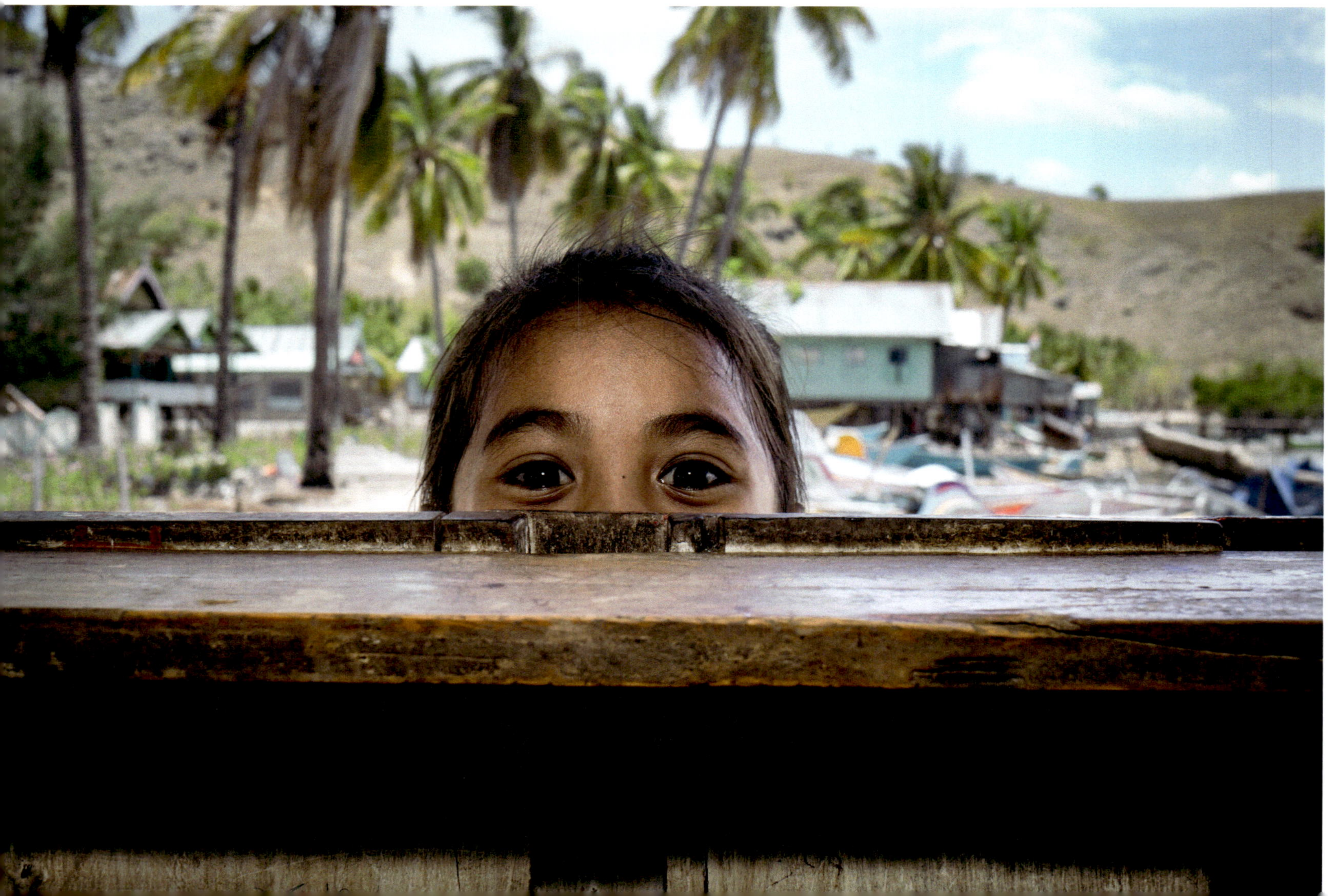

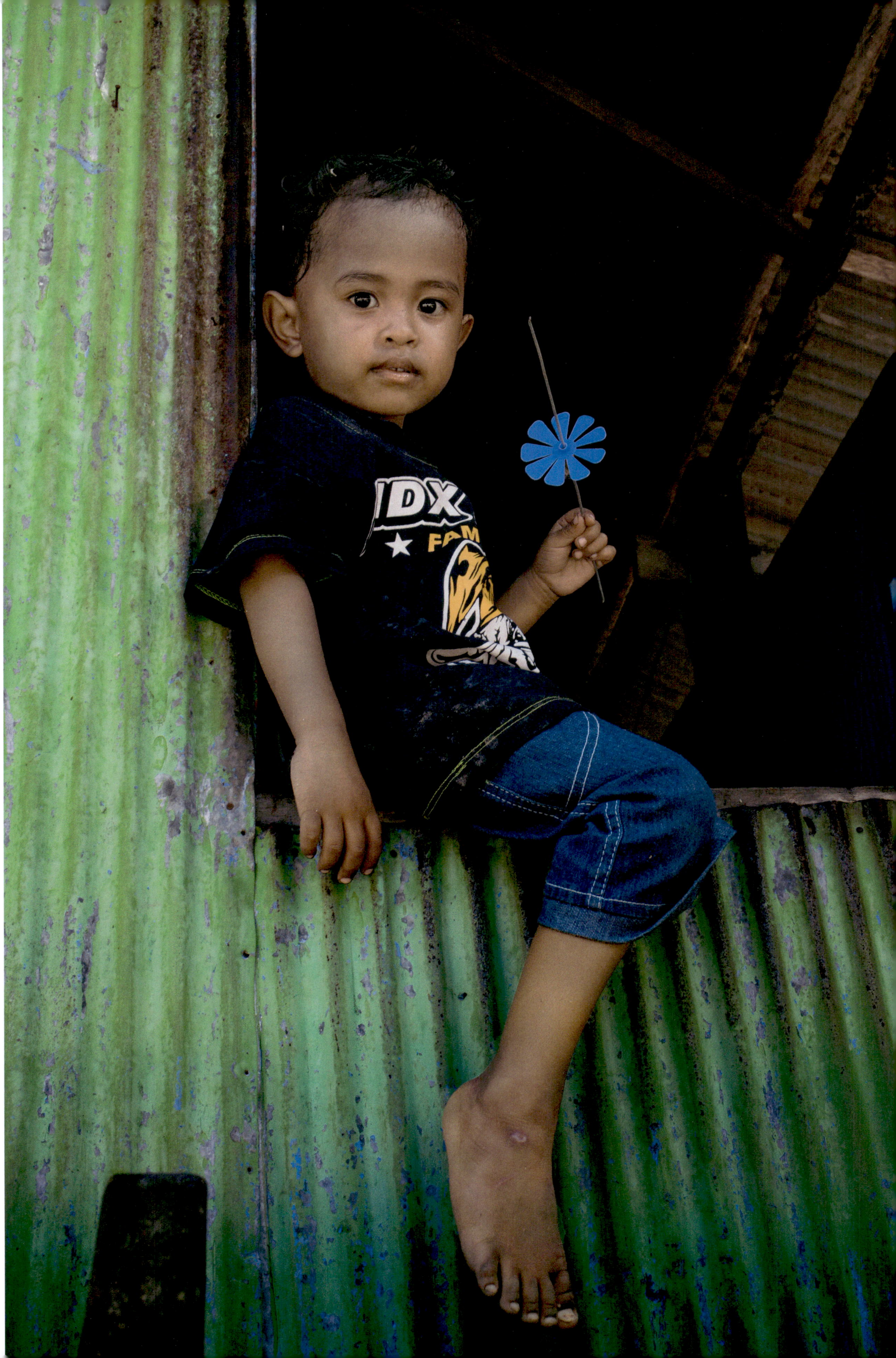

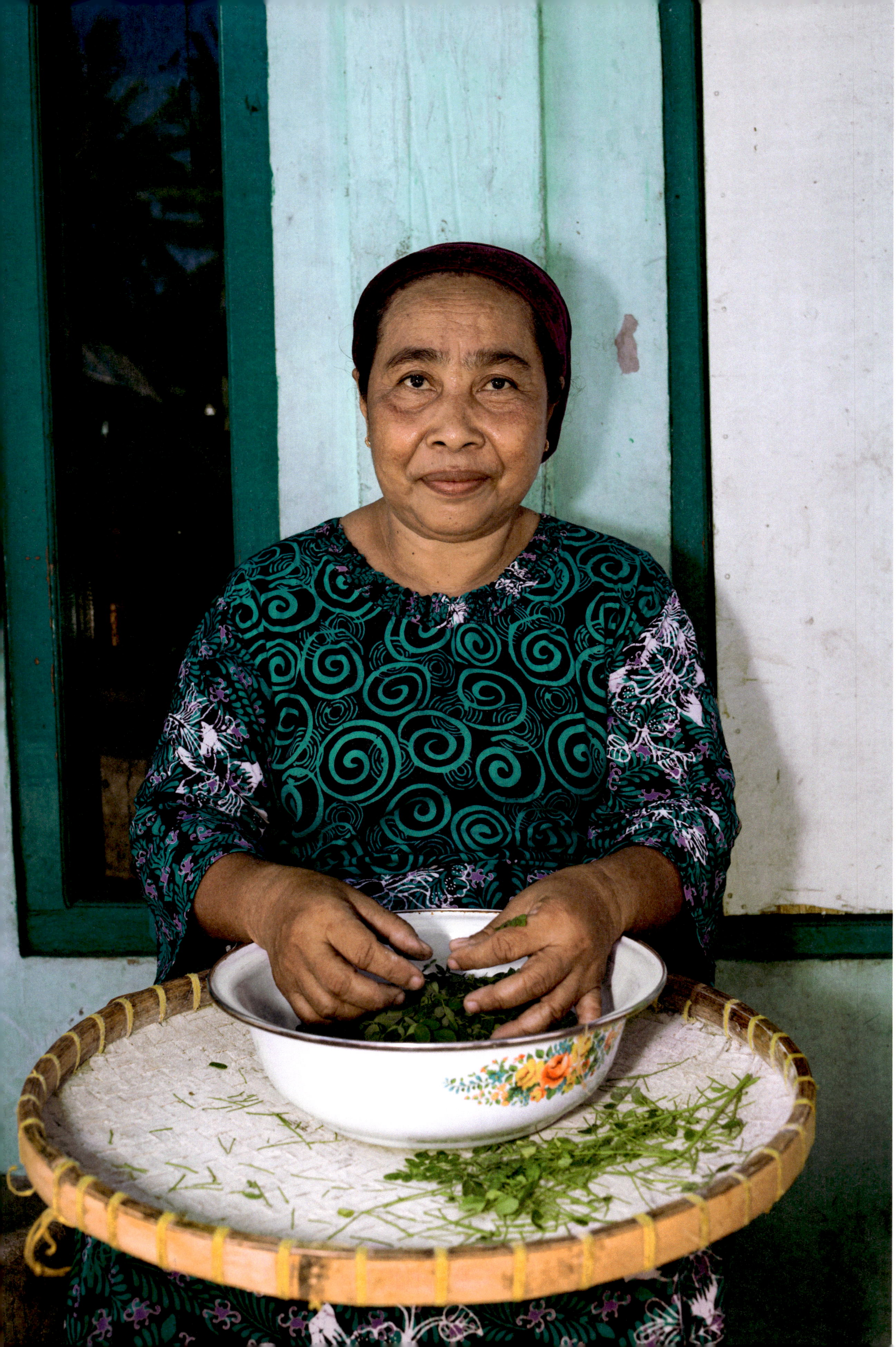

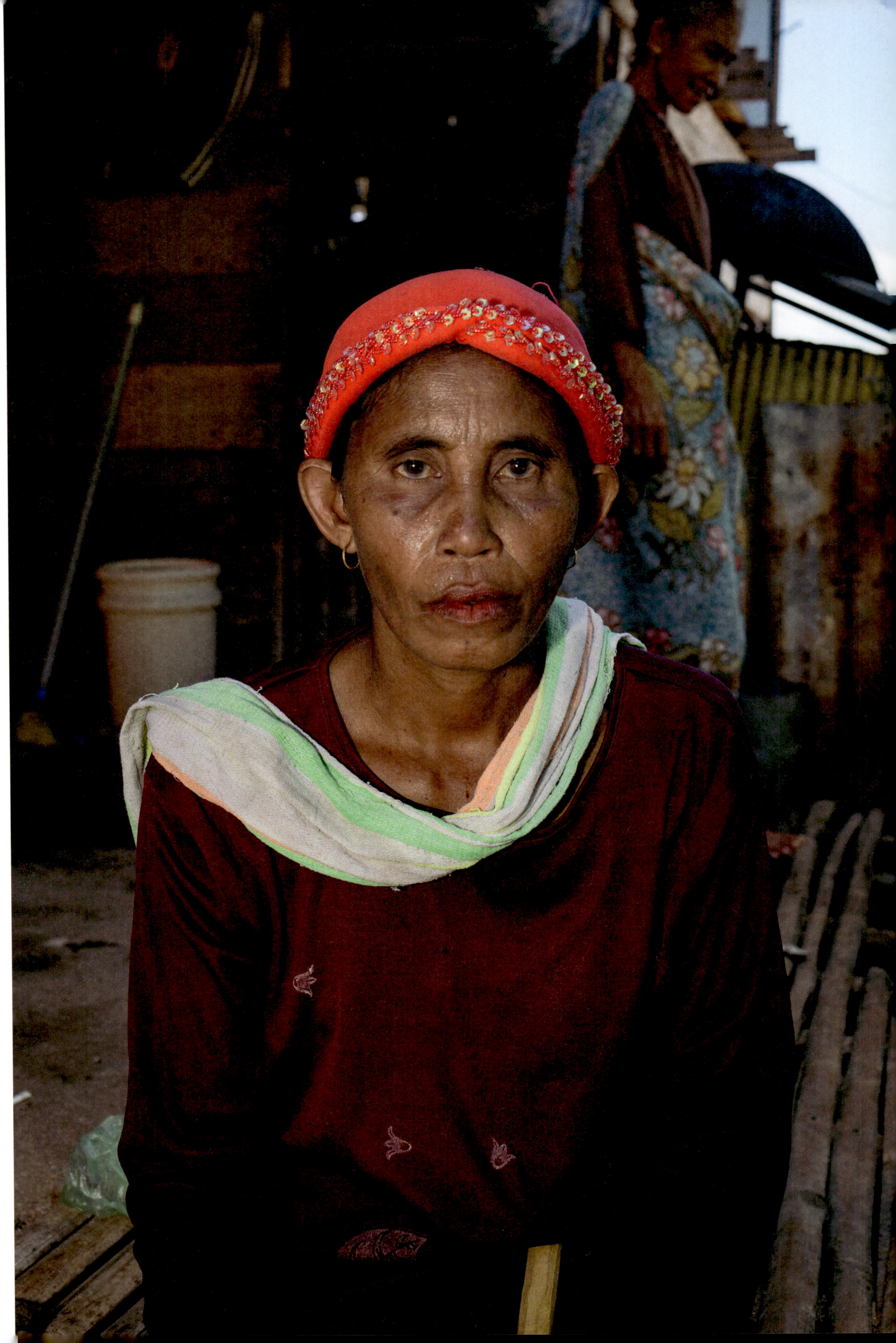

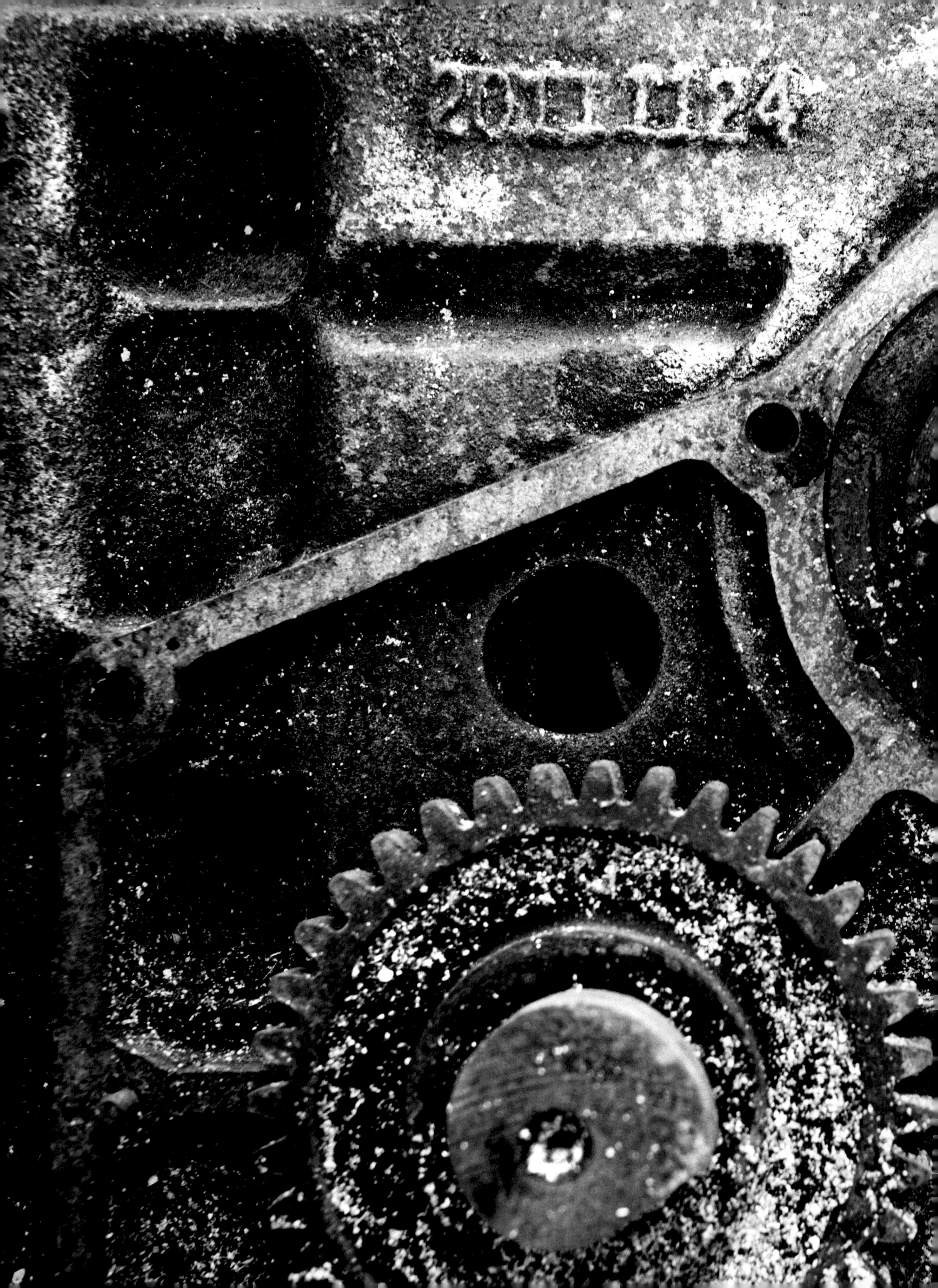

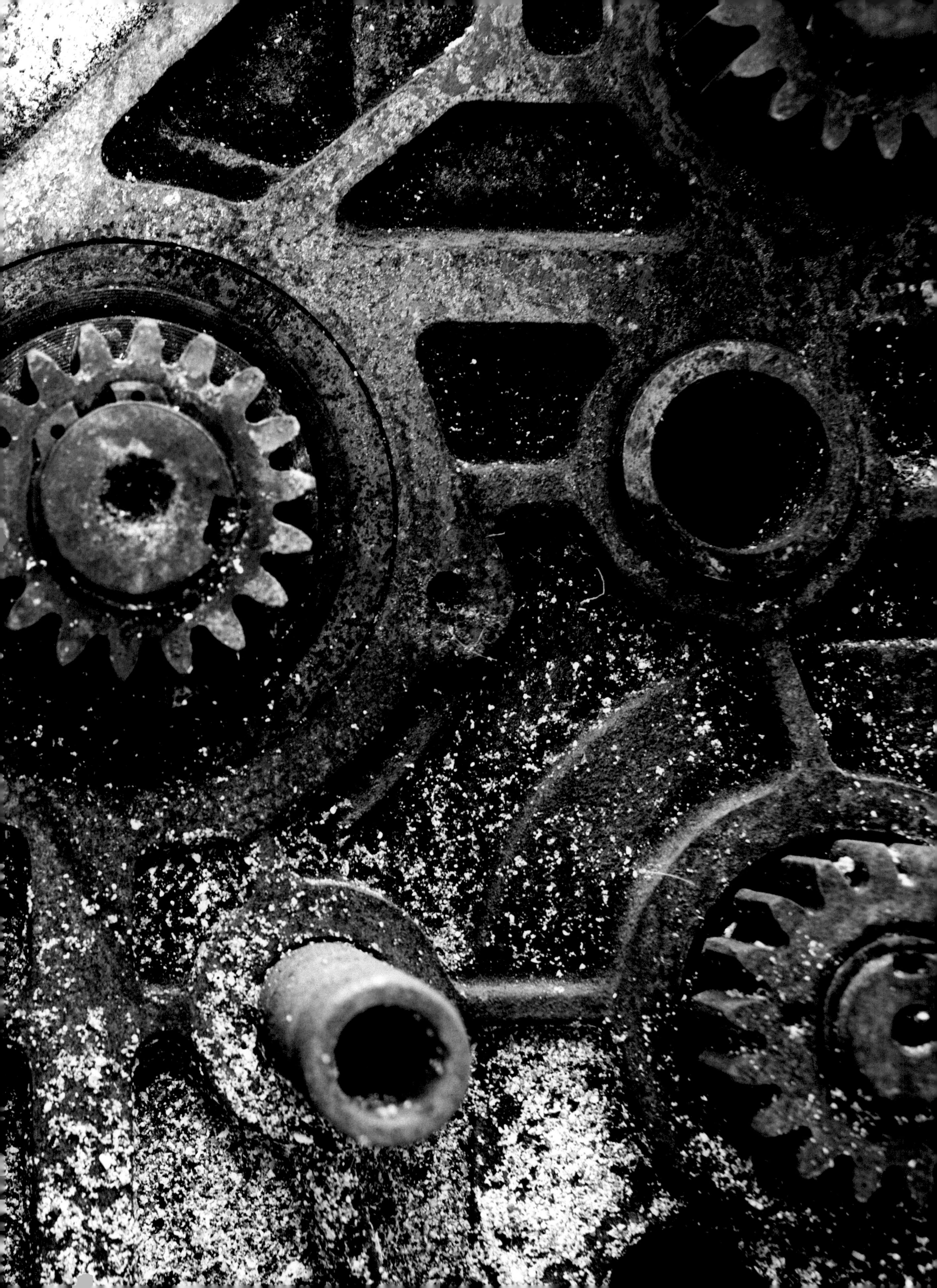

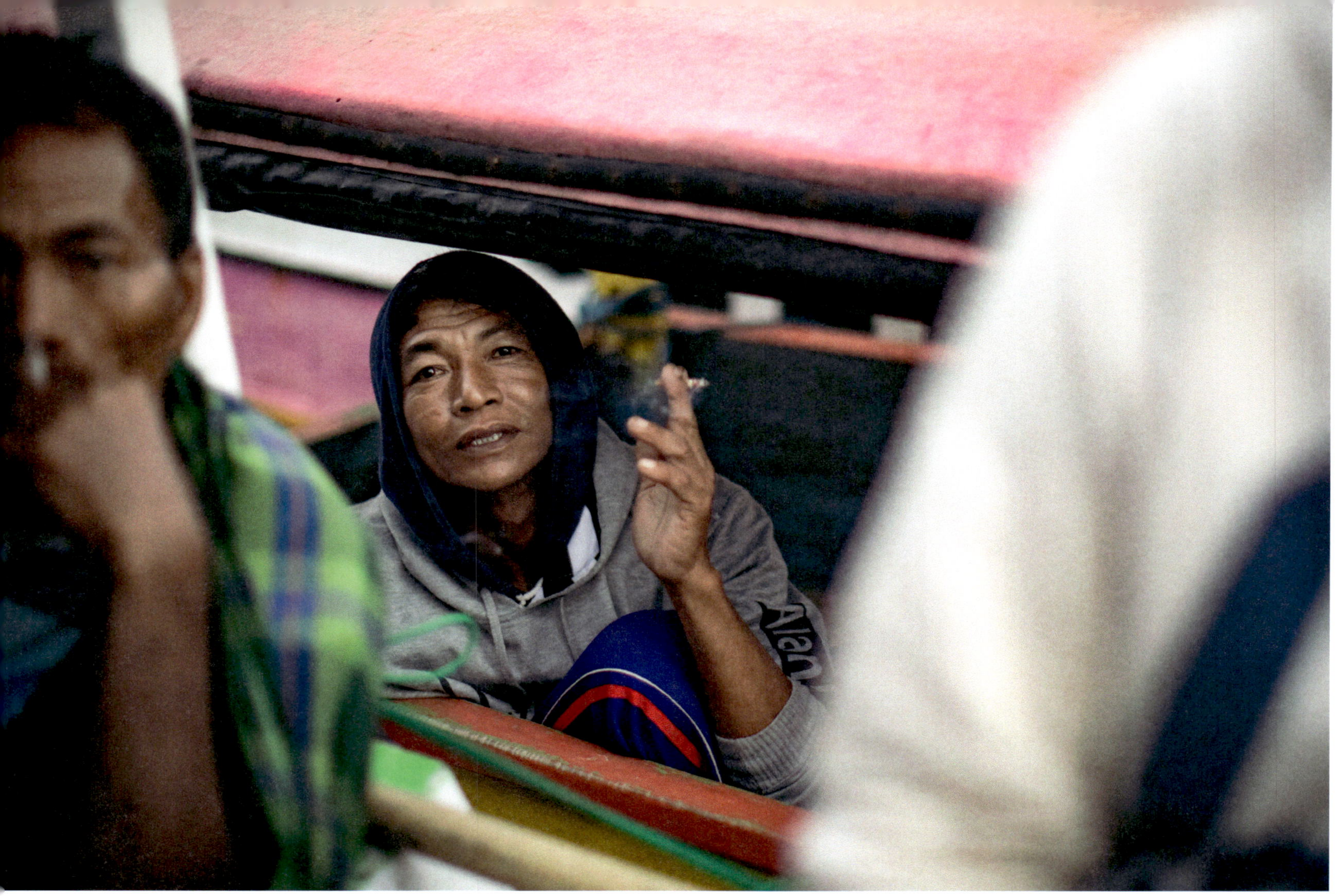

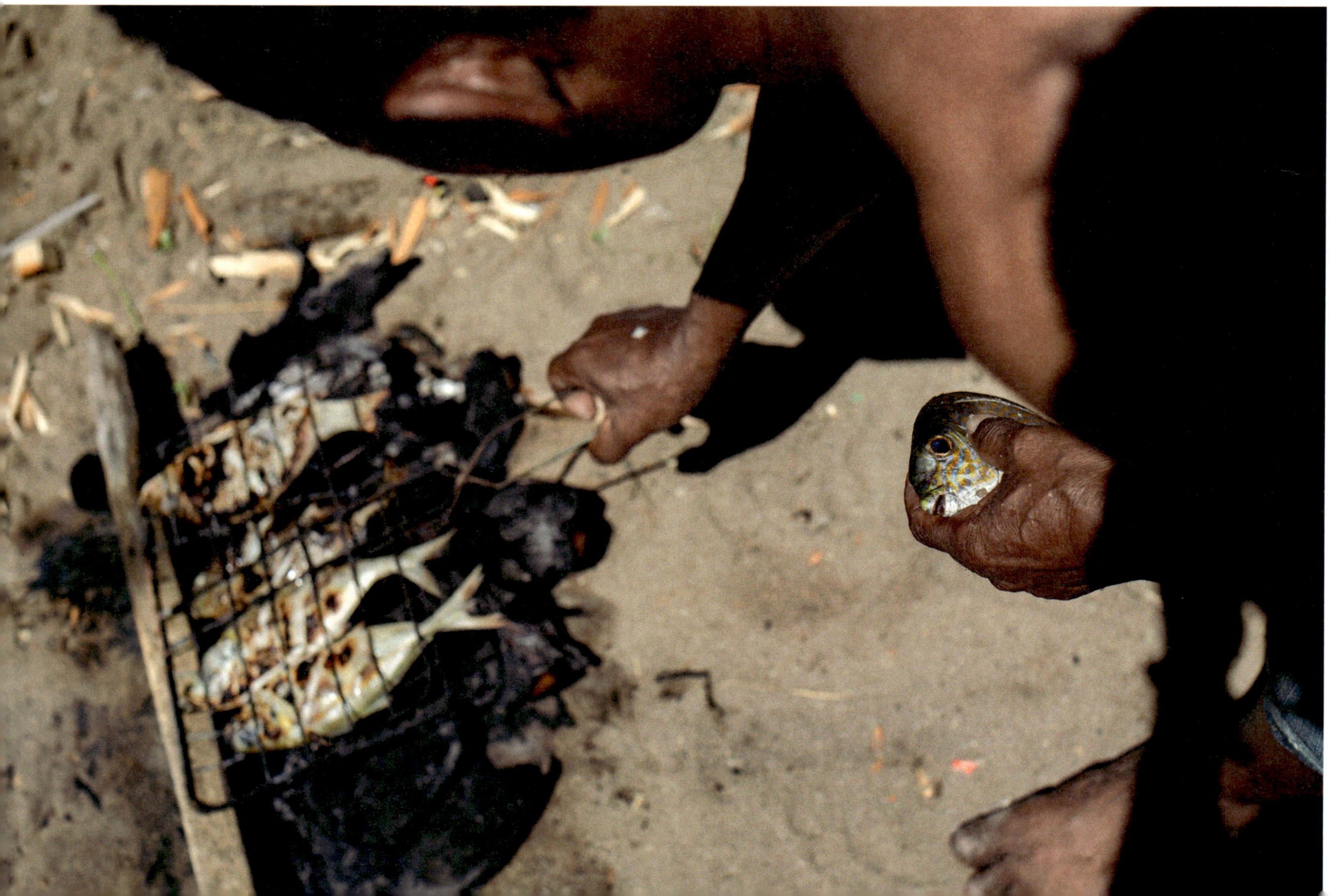

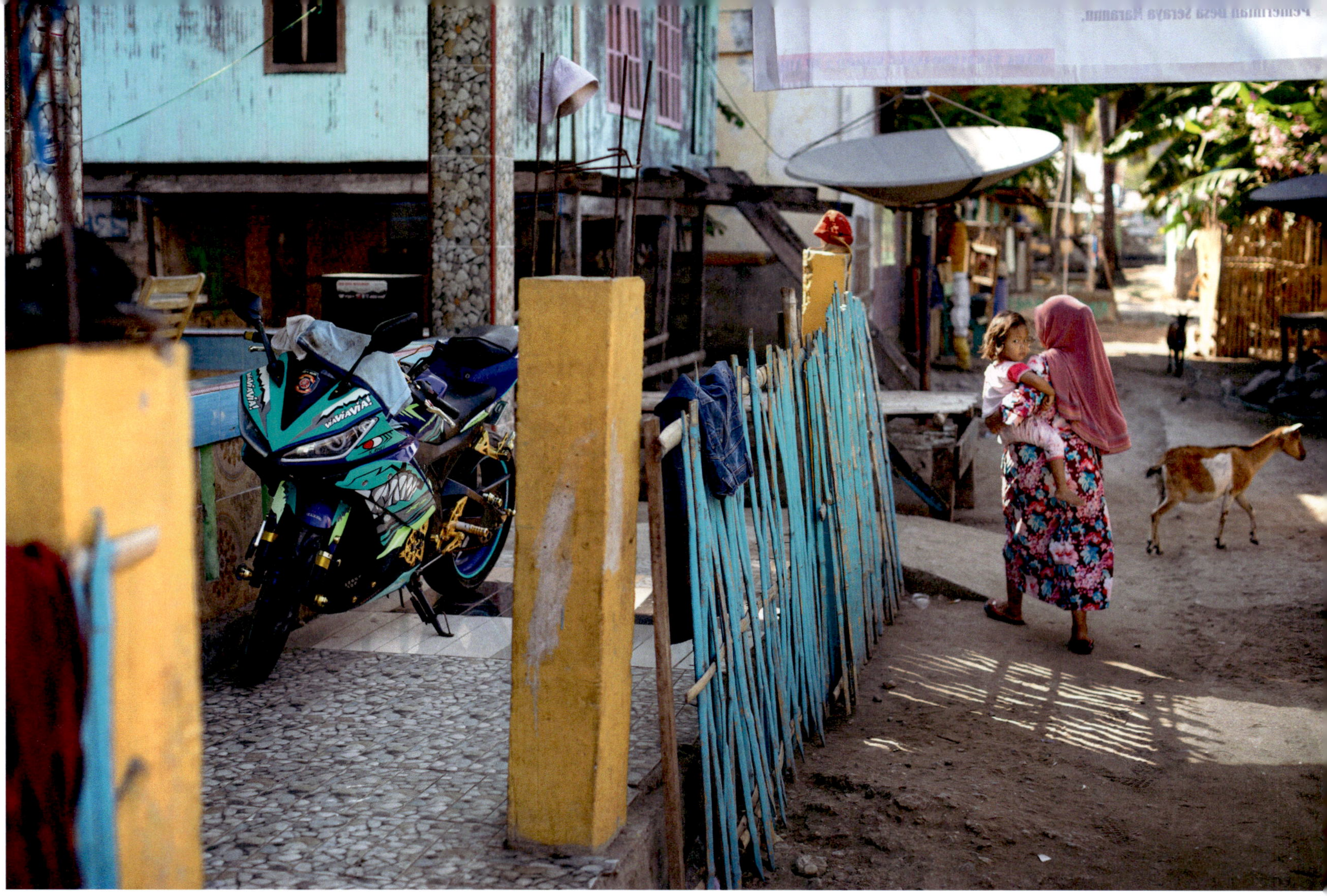

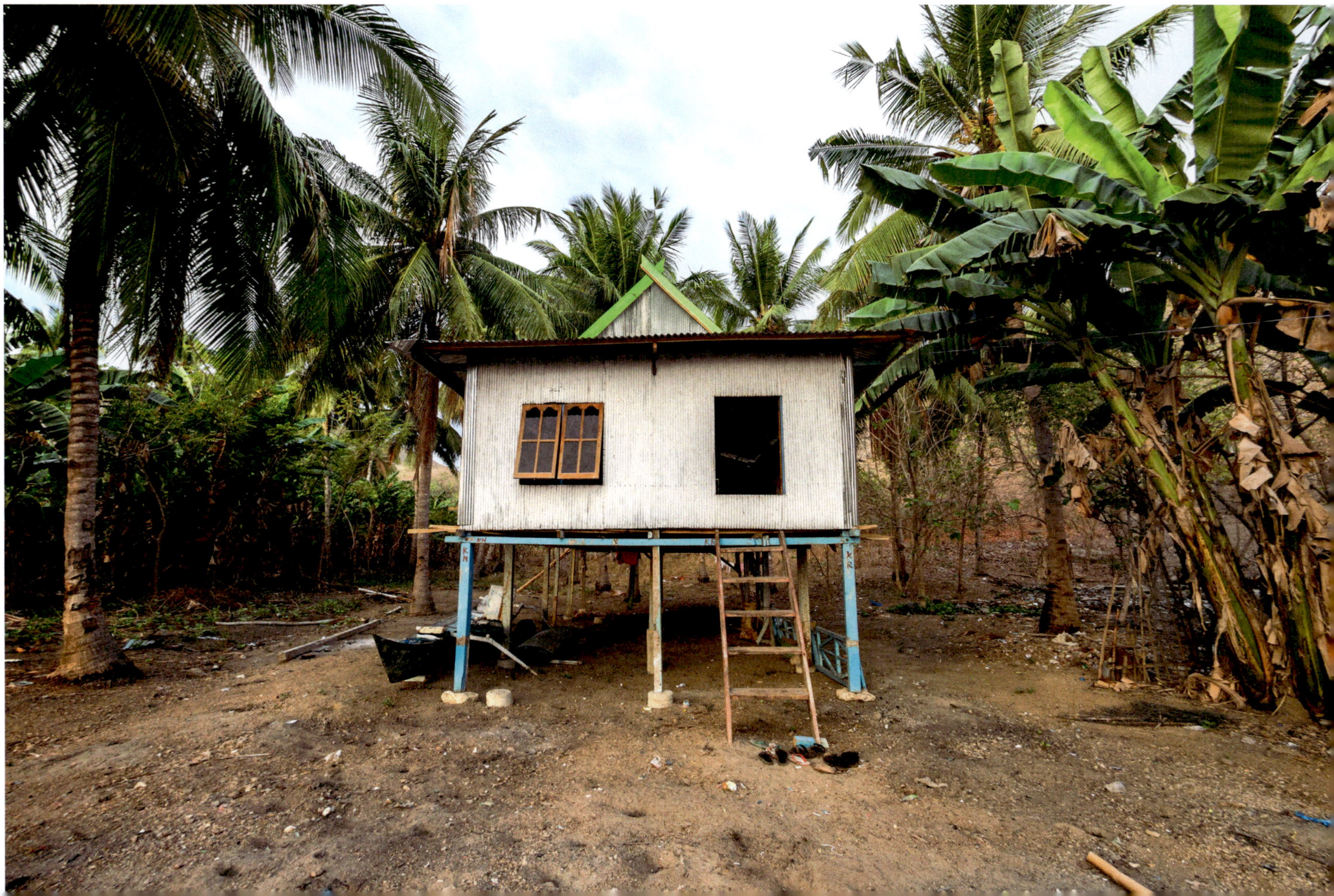

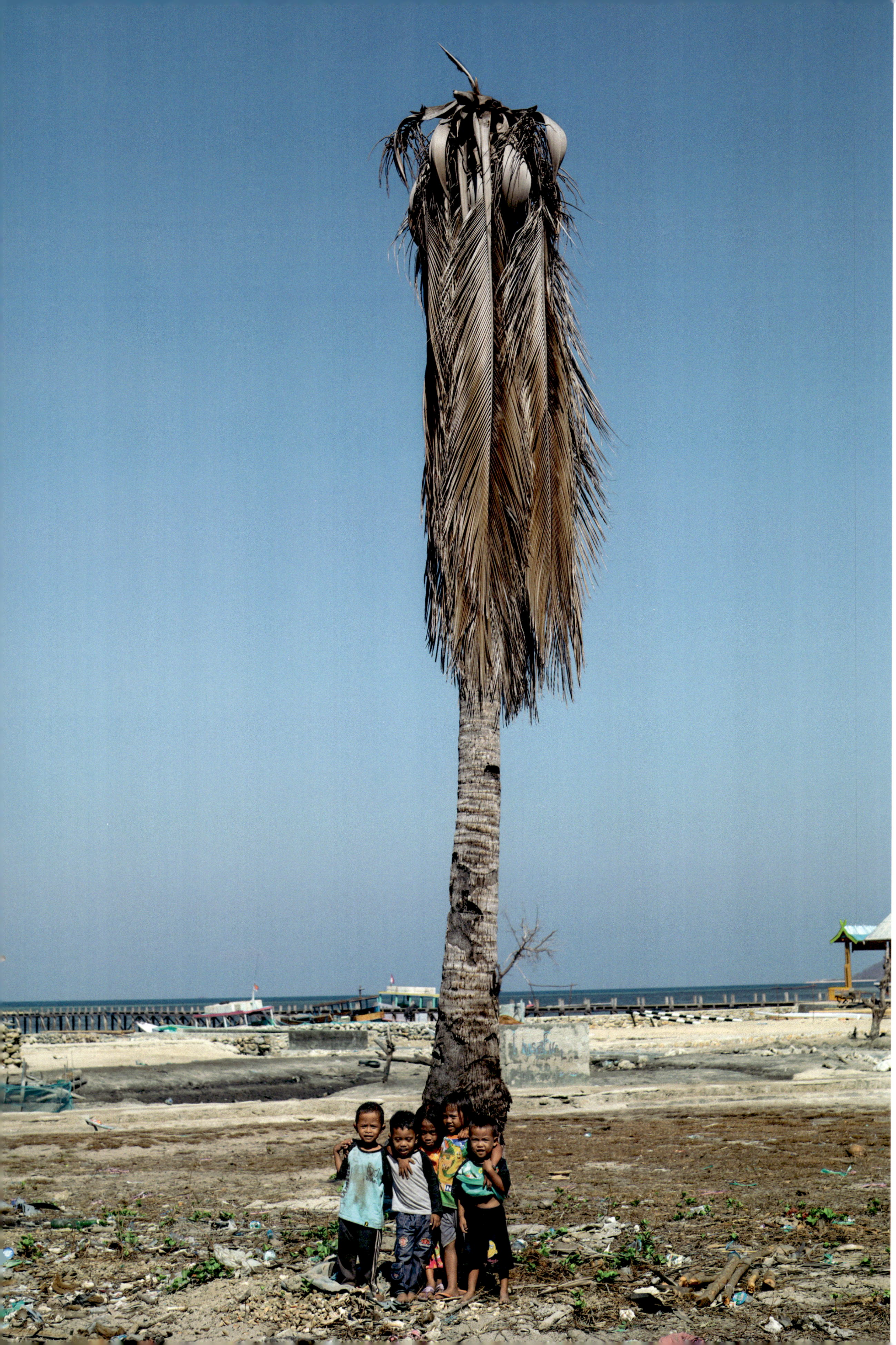

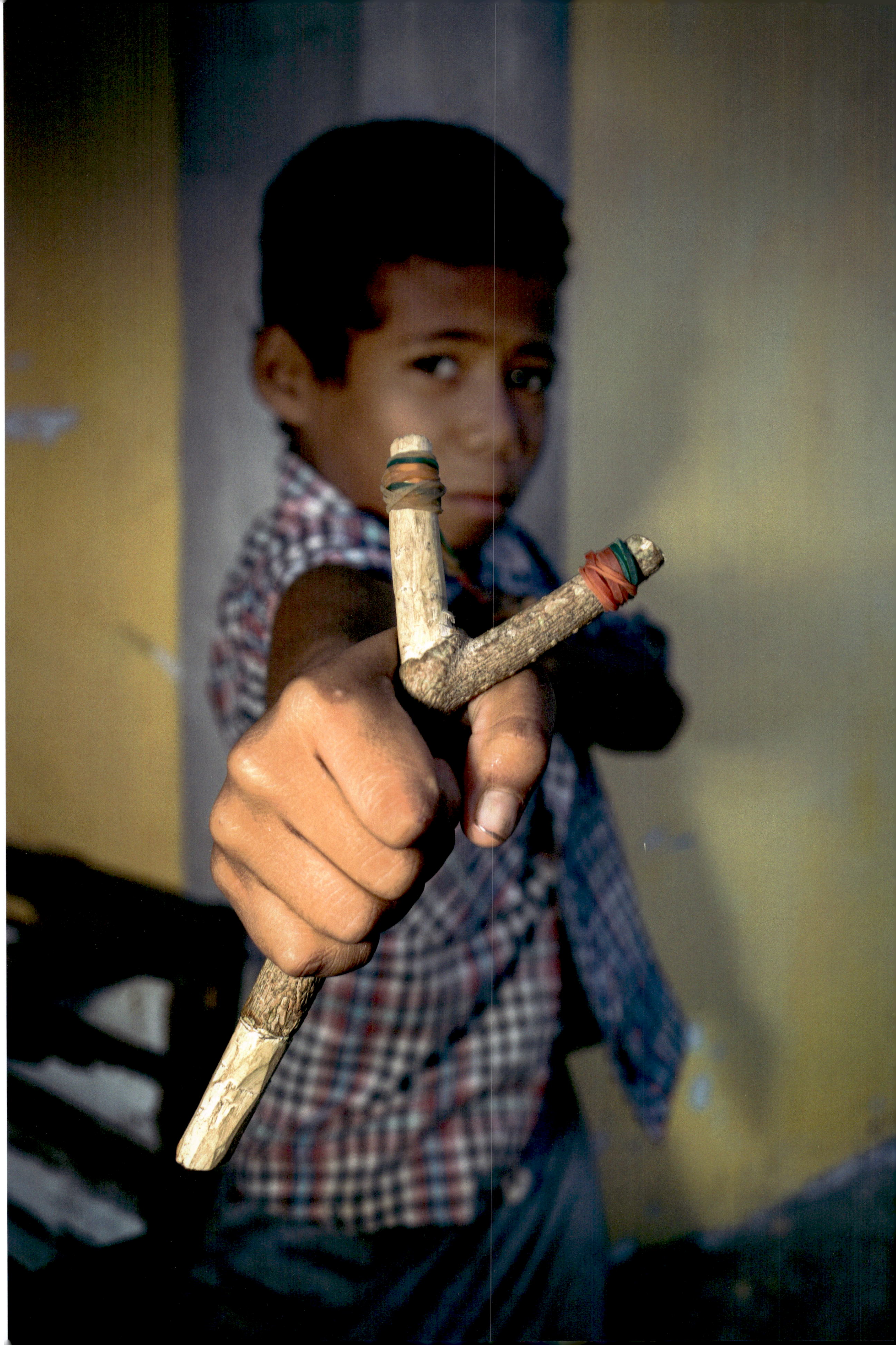

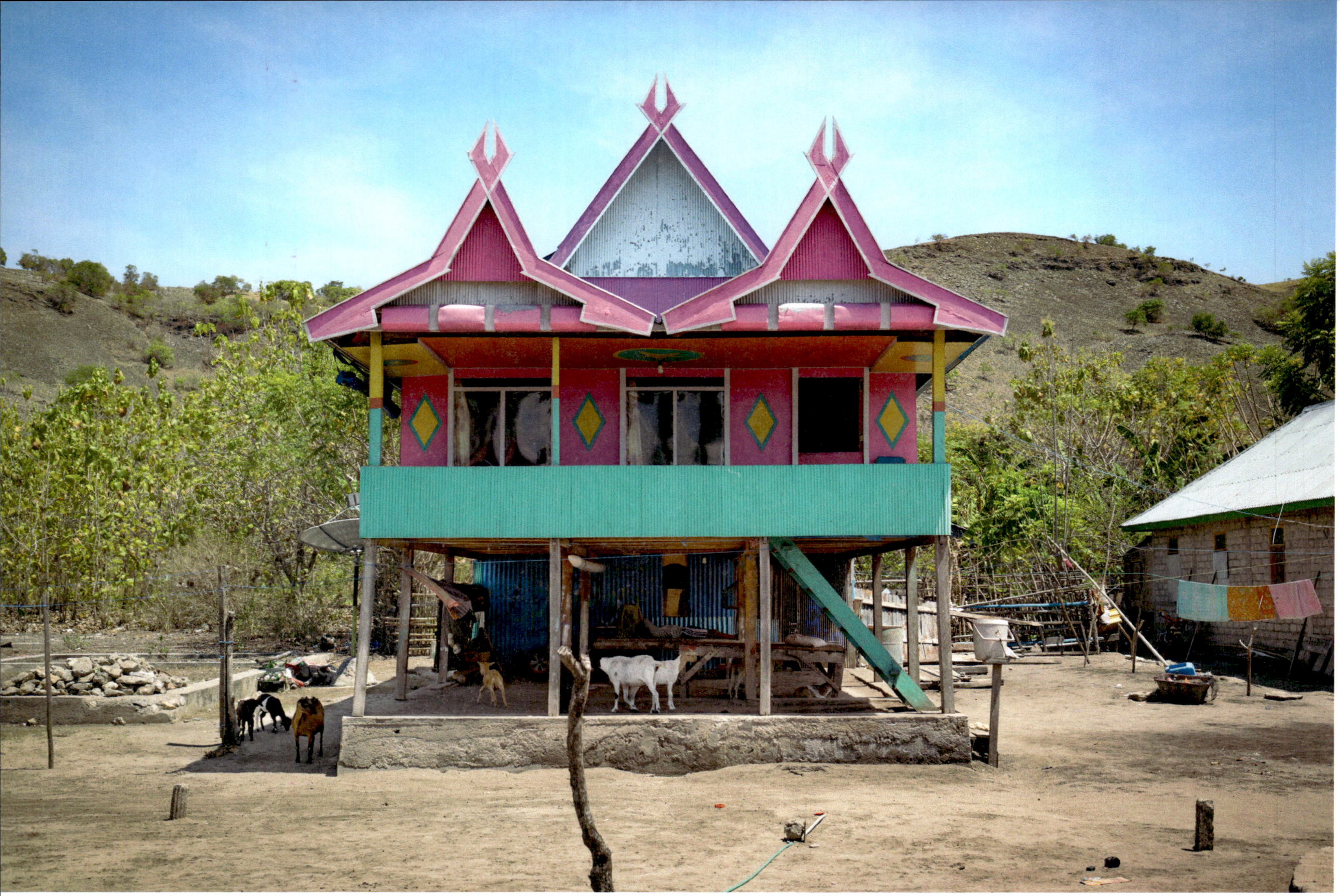

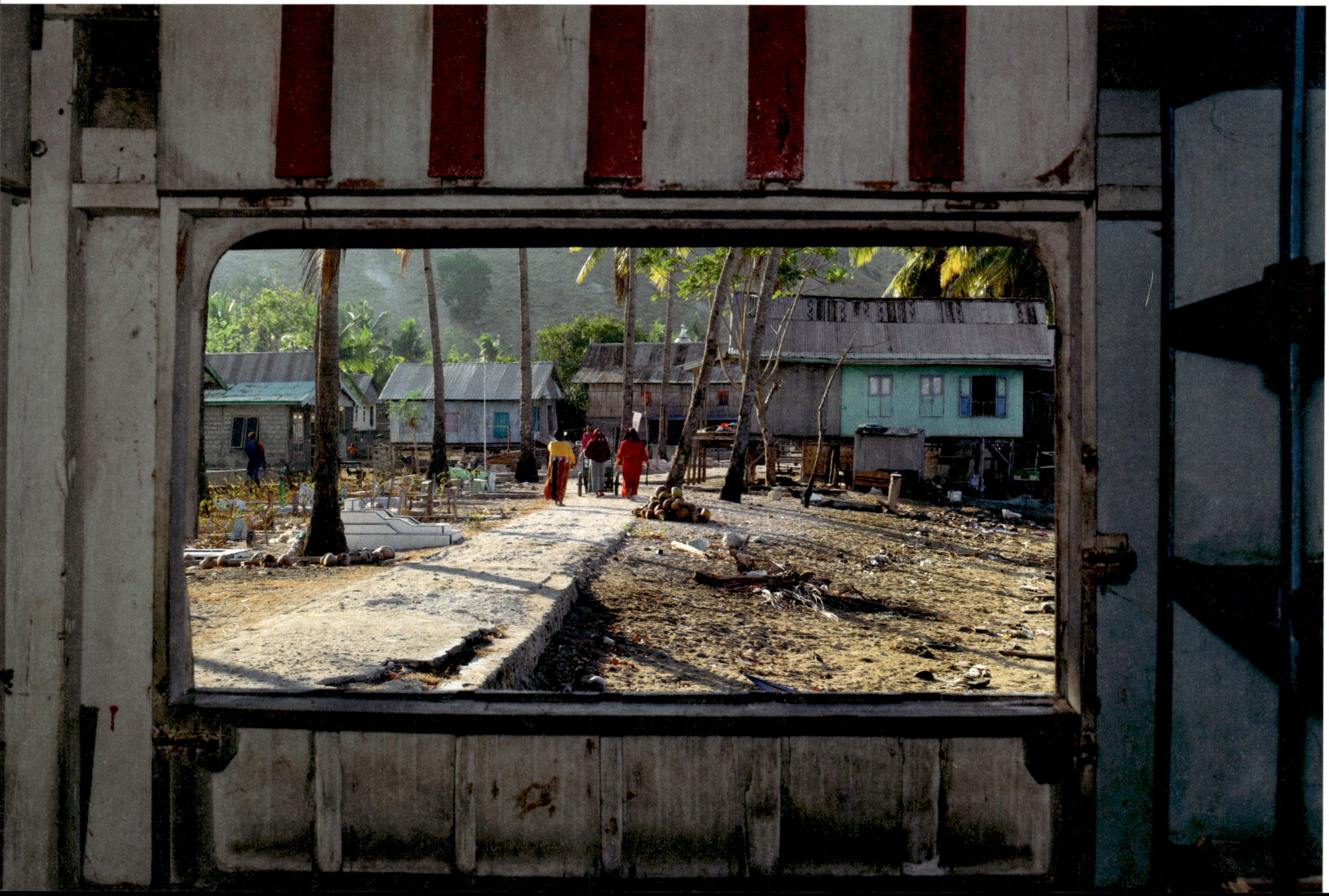

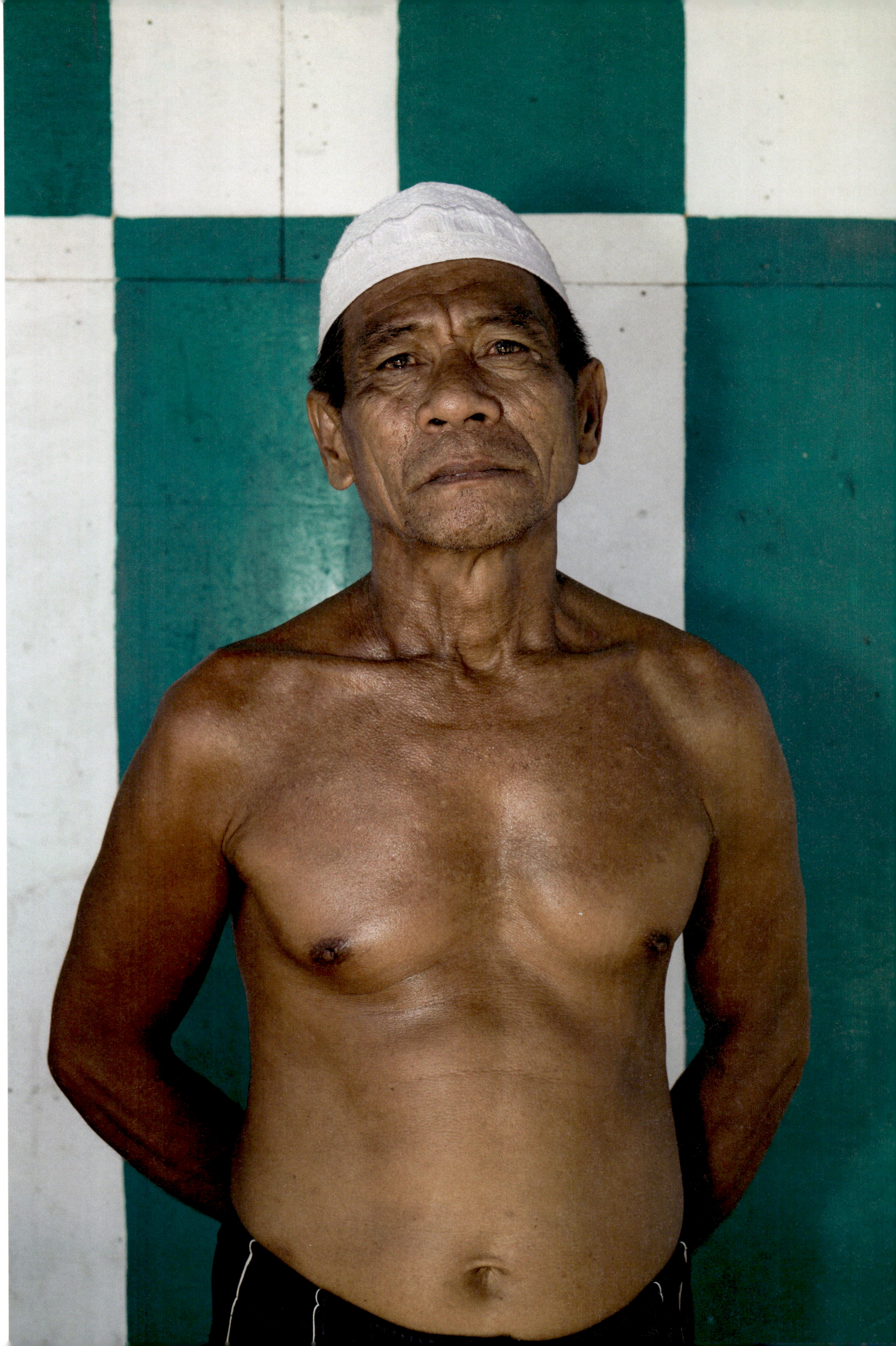

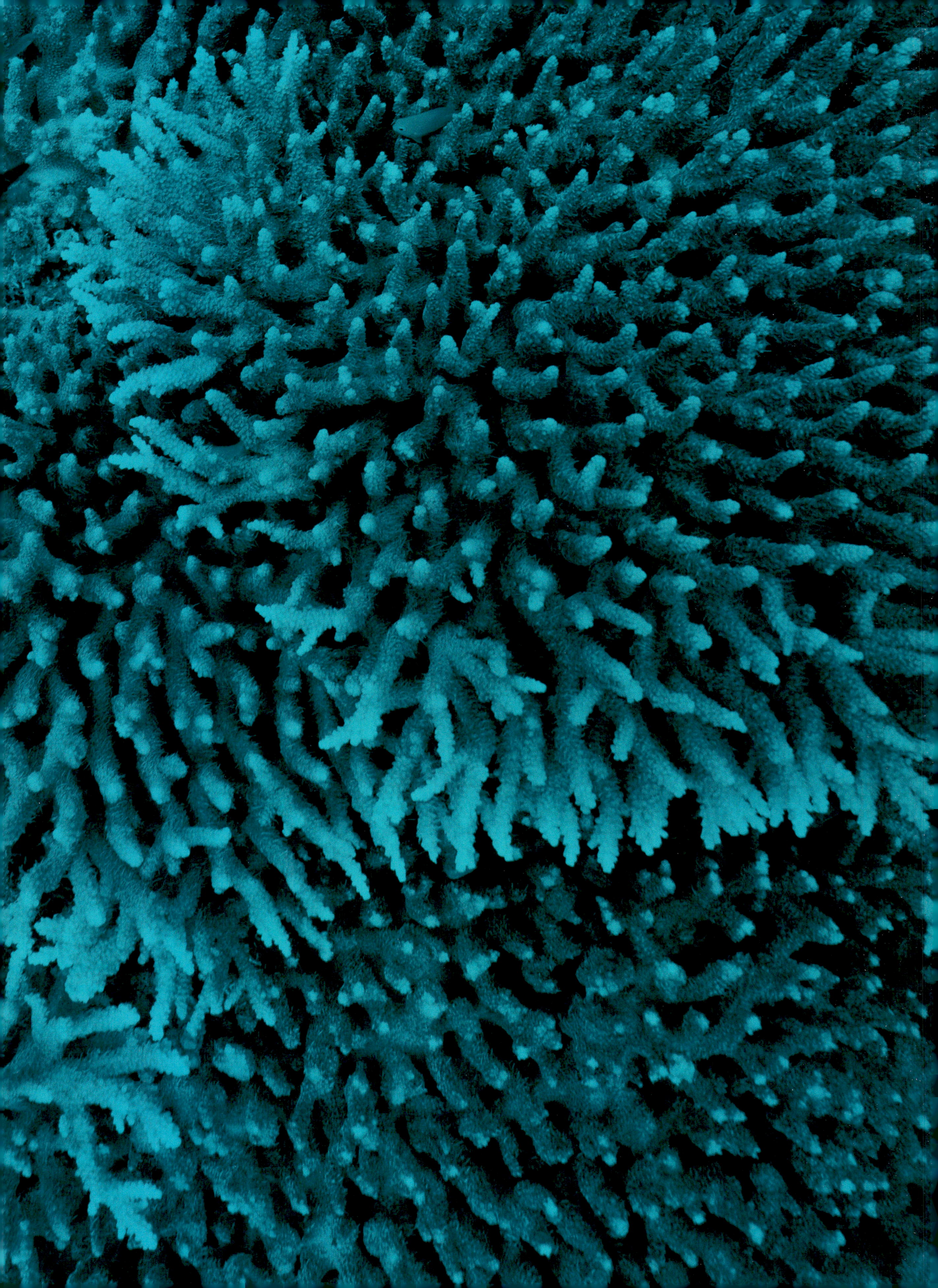

INTERWOVEN FATES

The sun is rising and the Ketinting are already out at sea off Seraya Besar. On these traditional boats, some small fish have been caught to use as bait for bigger fish – tuna, swordfish and jack mackerel. One person, or, with more difficulty, two, can sail these skiffs to go fishing along the coast. They didn't used to need motorised boats, but now they do, having to consume fossil fuels in order to go further out so they can get enough for the village to eat every day. Dynamite hasn't been used in fishing here for more than 10 years, but it has left its mark, forming a desert-like underwater landscape around the island where nothing can grow and forcing fishermen to go further and further out to get a catch. Fishing expeditions that go further mean longer days and greater risks taken in order to earn enough from their job, one of the most dangerous in the world.

Martin tagged along on these fishing trips. The fishermen got used to seeing him, on Komang's or Dulah's boat or even freediving. They slowly began to accept his presence, noticing that he didn't scare off the fish. Now, they call him over whenever there's a good catch, to immortalise their haul of fish. Martin photographs them stationed on the surface of the water, just above the reef, so close to the coral out of which they're drawing fish like water out of a well. In harmony. His photos show their ingenuity, their precision when angling or fishing with a net. Respectful fishing, only seeking out what's necessary, leaving the rest of the underwater landscape in peace, like the young shark with black-tipped fins, searching for food and shelter in the forest of coral.

The days are long, intense, adventurous. Regularly punctuated by a struggle with the animal that can last for hours, on the end of the line, with their bare hands, with a roll of bamboo to lift the fish on board. Struggles that are sometimes won, sometimes lost. Their families and children wait impatiently, curiously, for their return home to dry land. The men's smiles are a sign of their pride after a successful day of fishing.

The fundamental link between coral and human hangs by a thread. The reef is generous and offers its food to men and women when they respect it in return. But, if the coral disappeared, where would humans go?

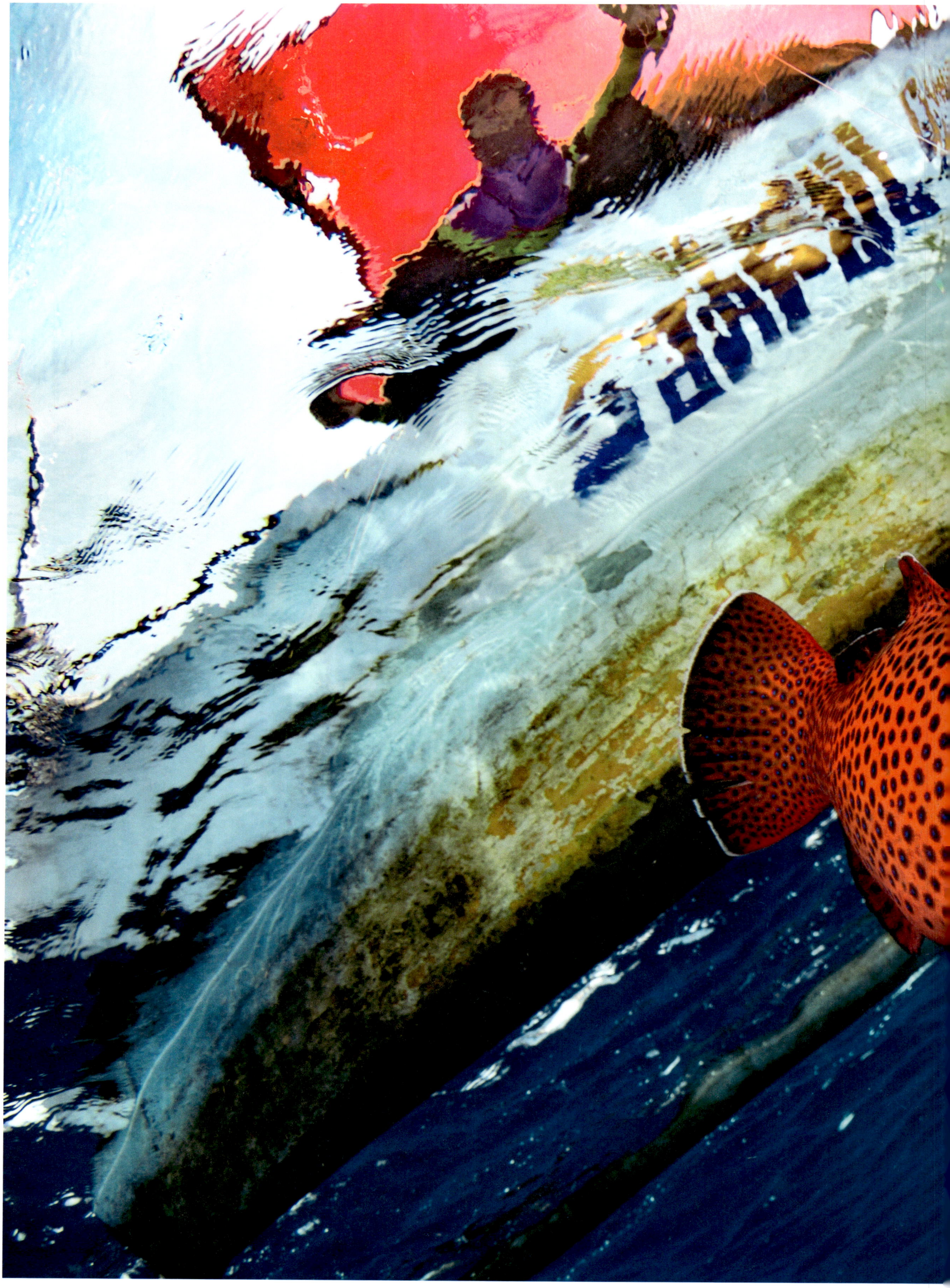

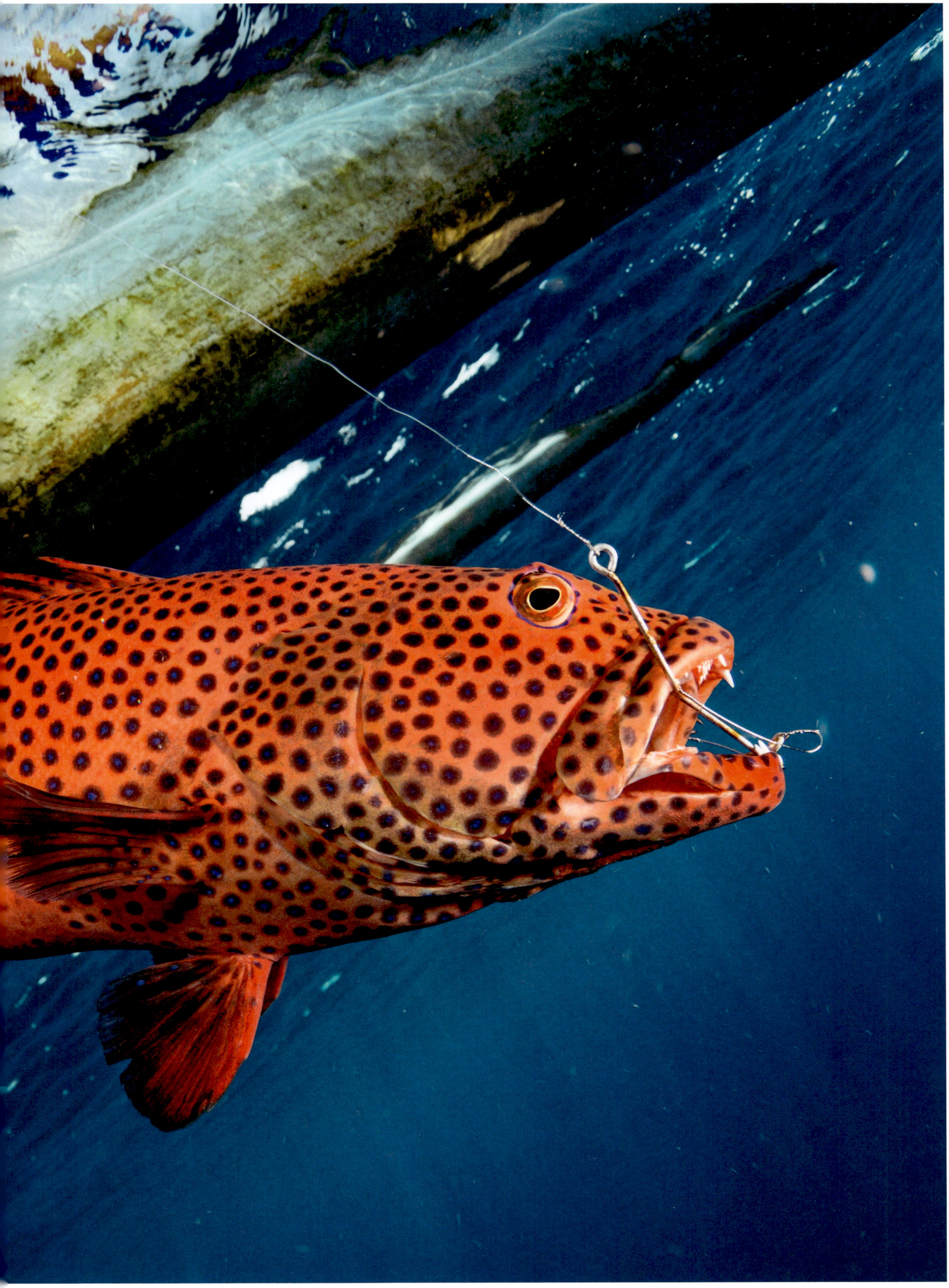

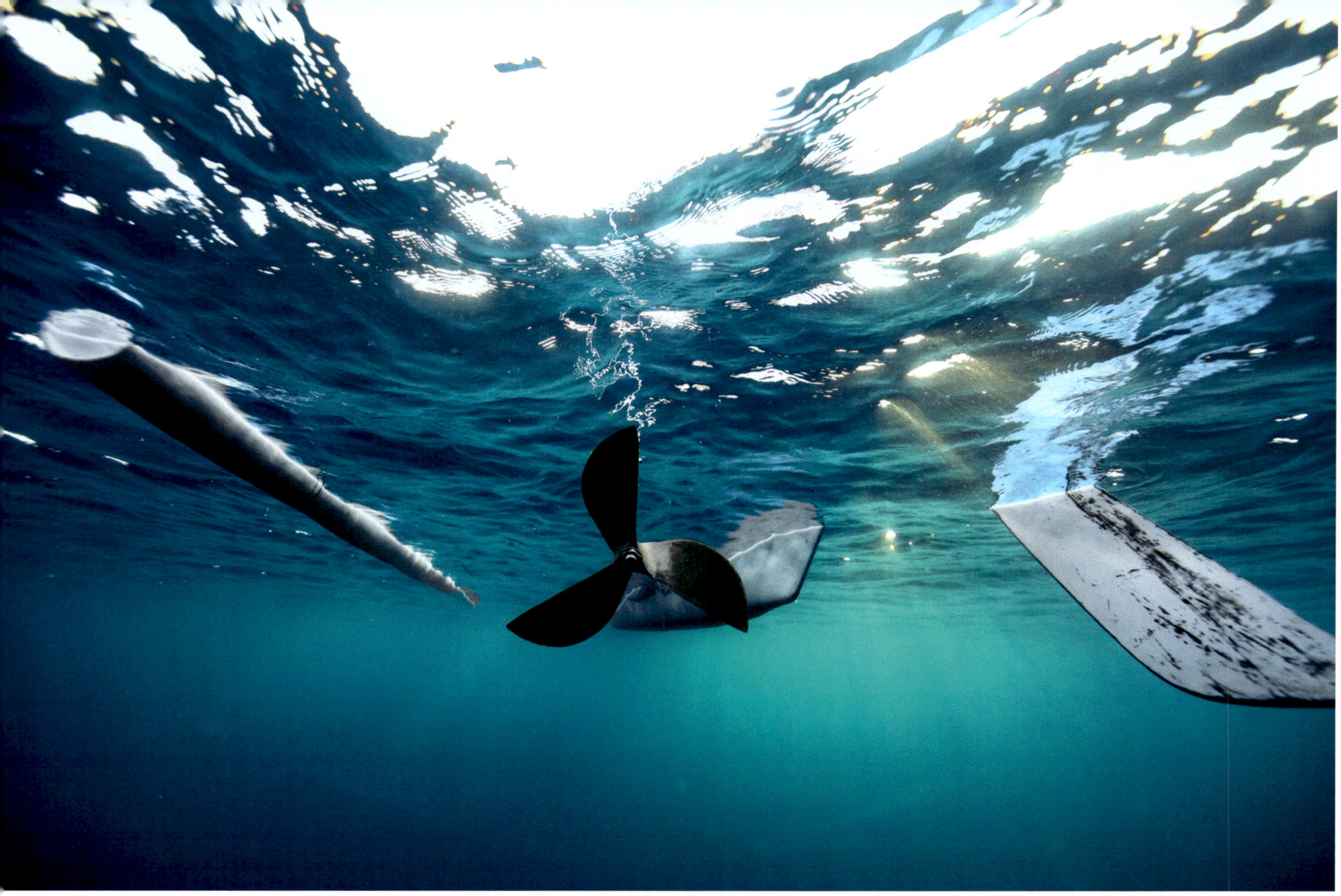

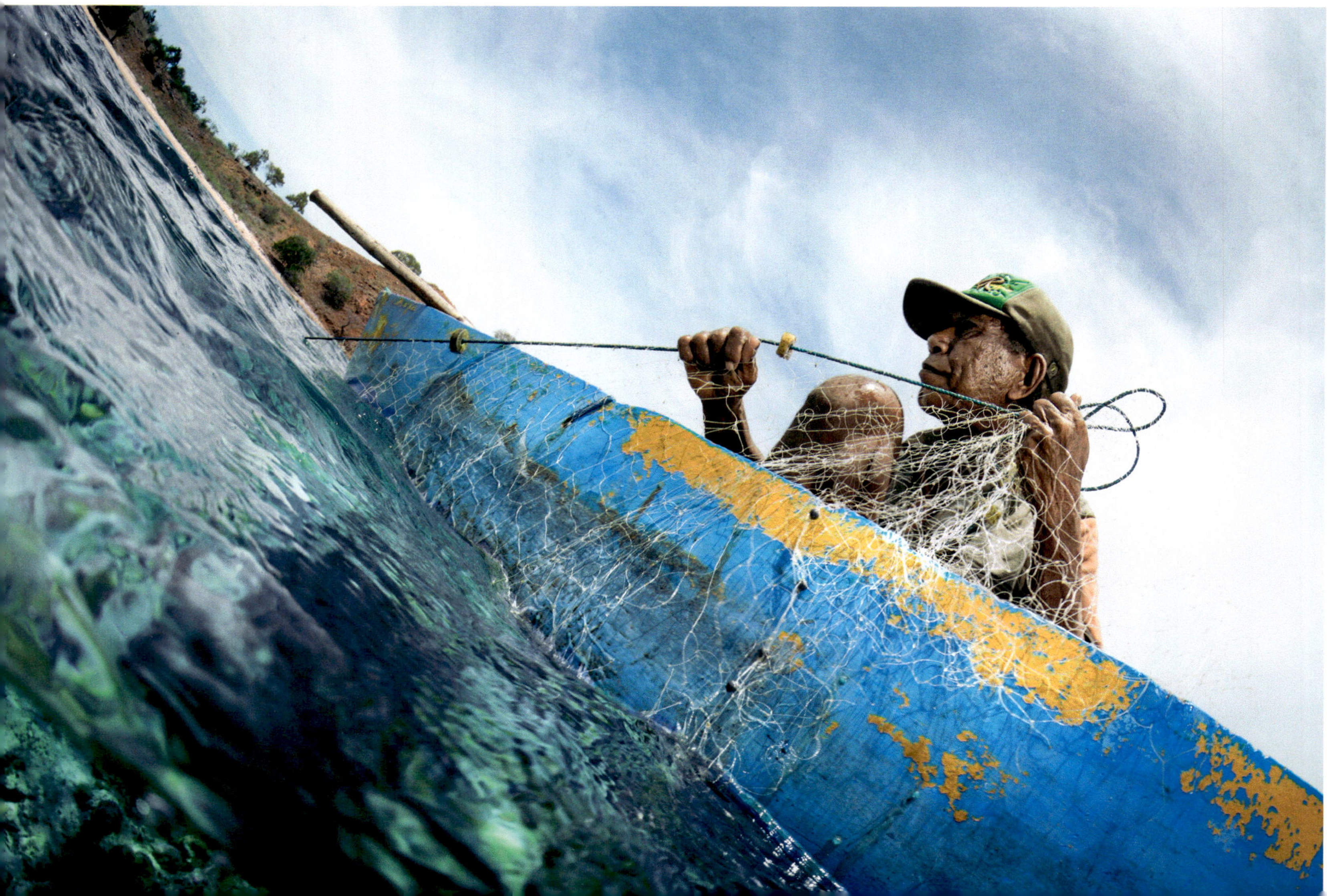

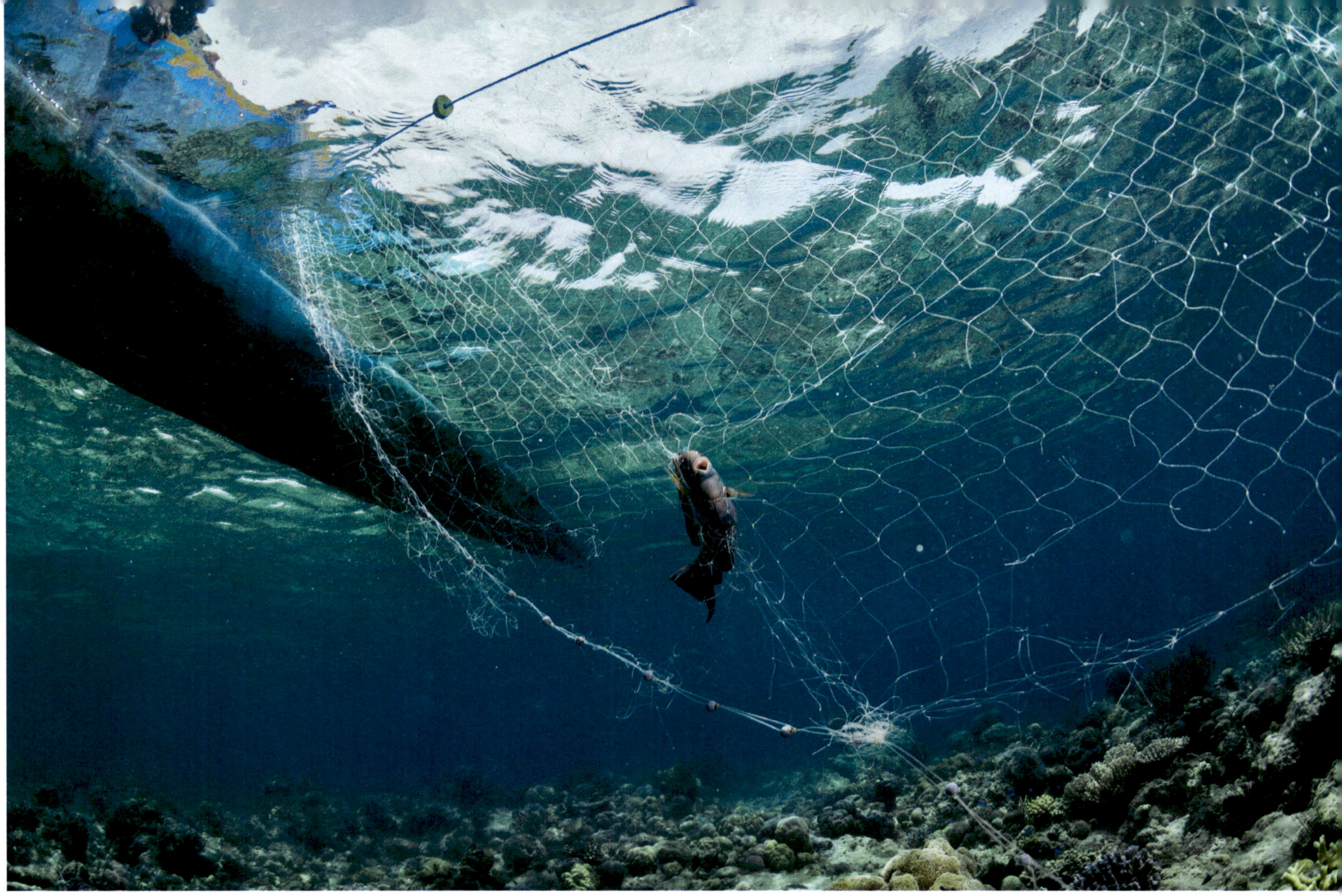

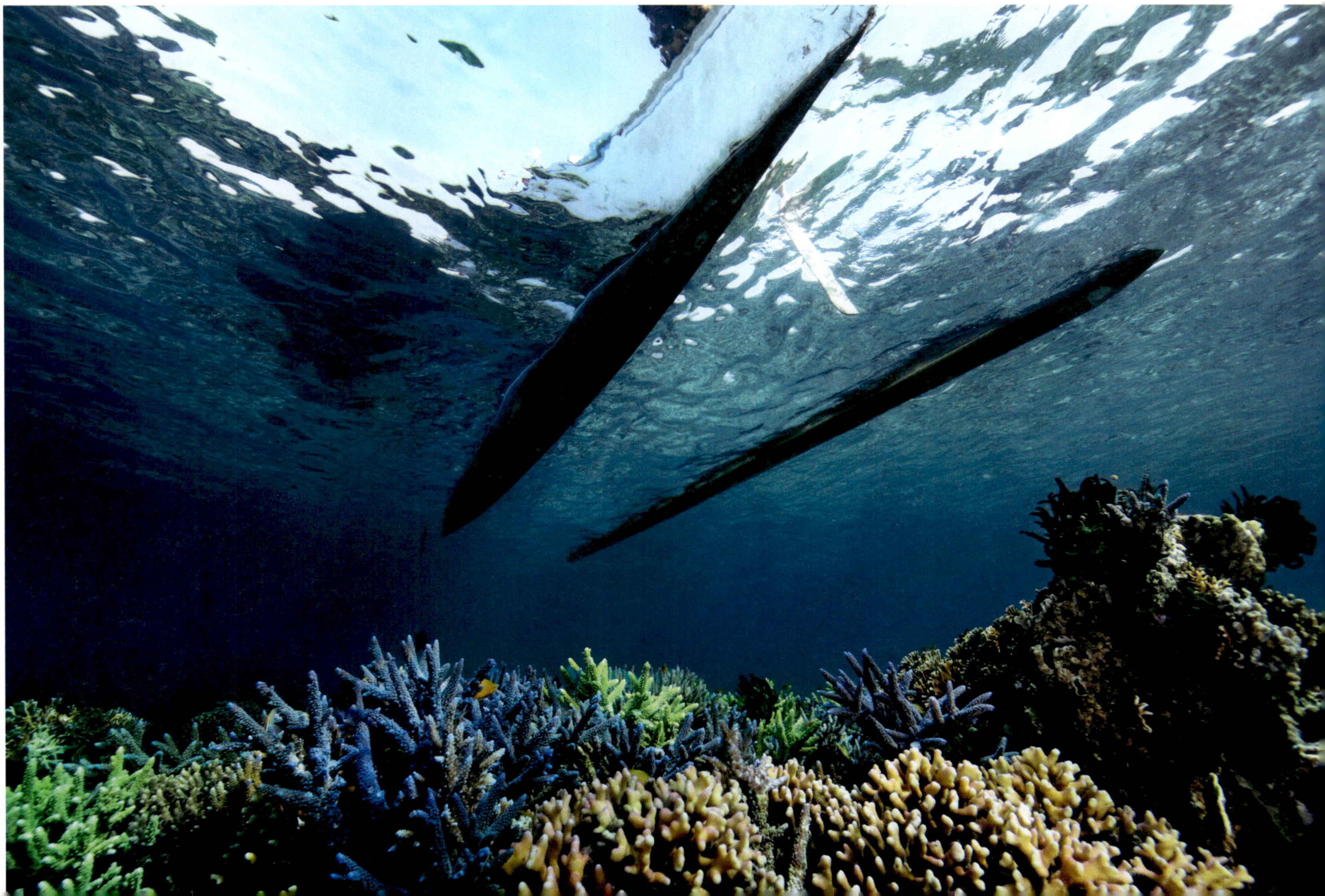

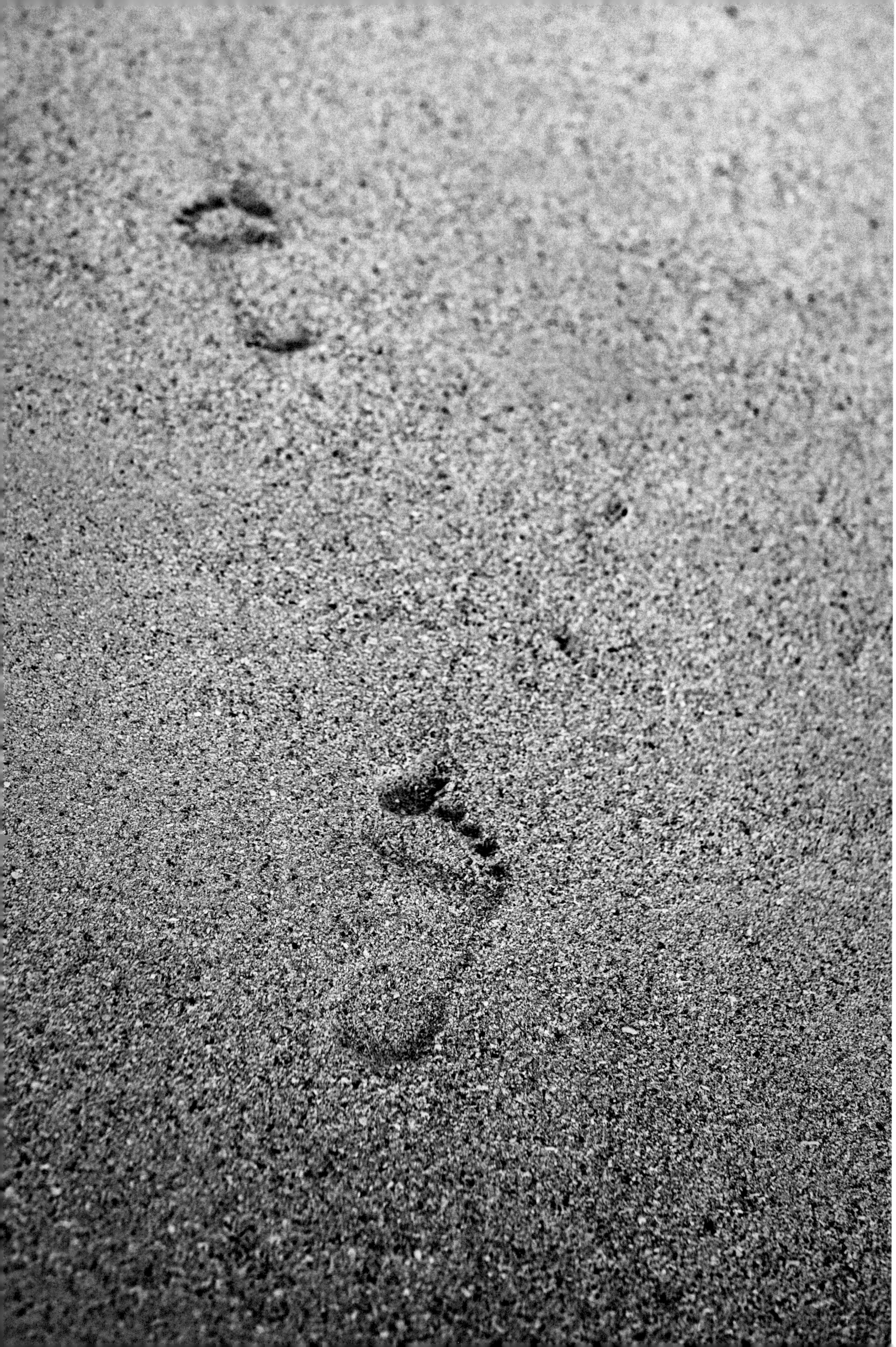

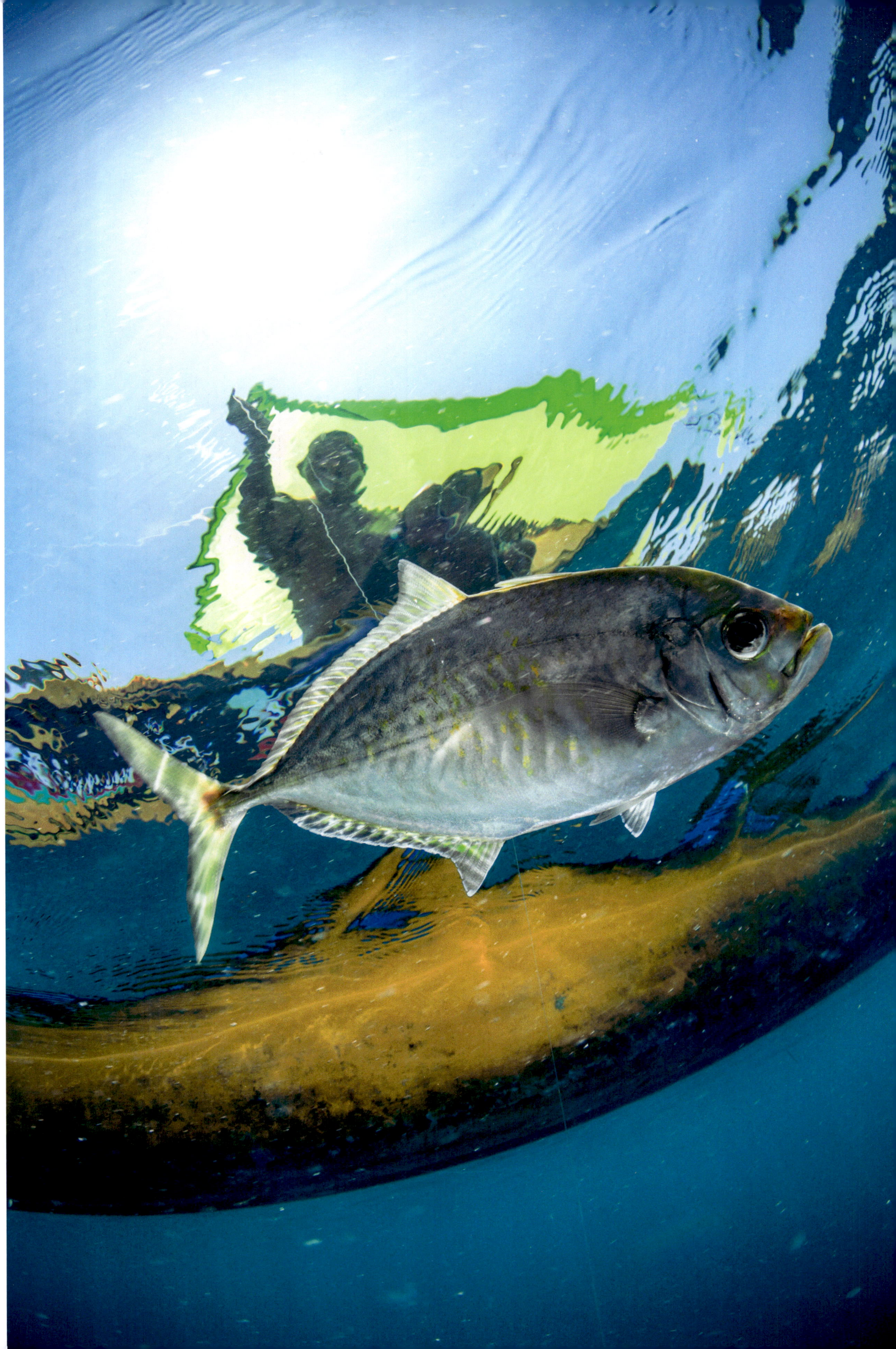

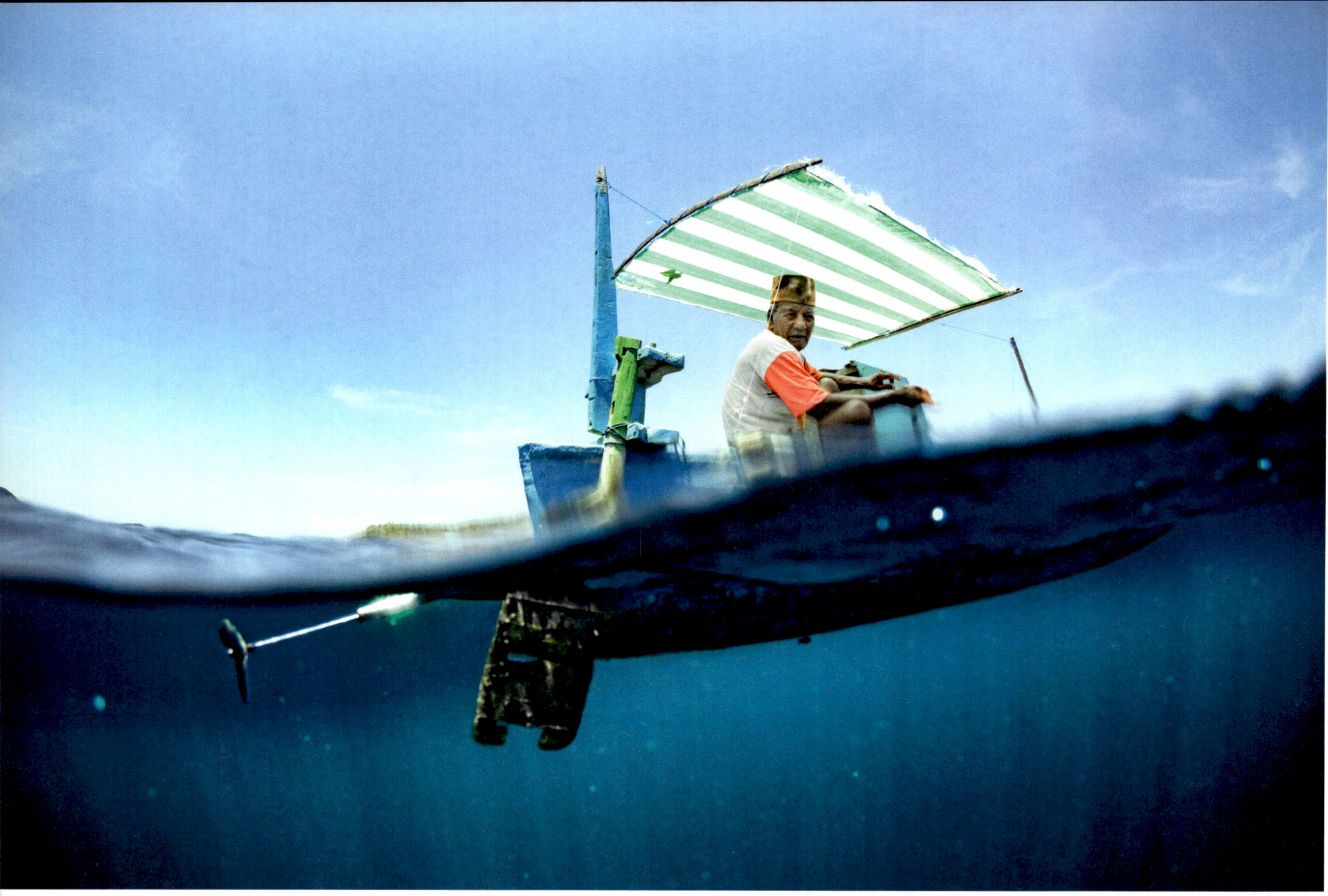

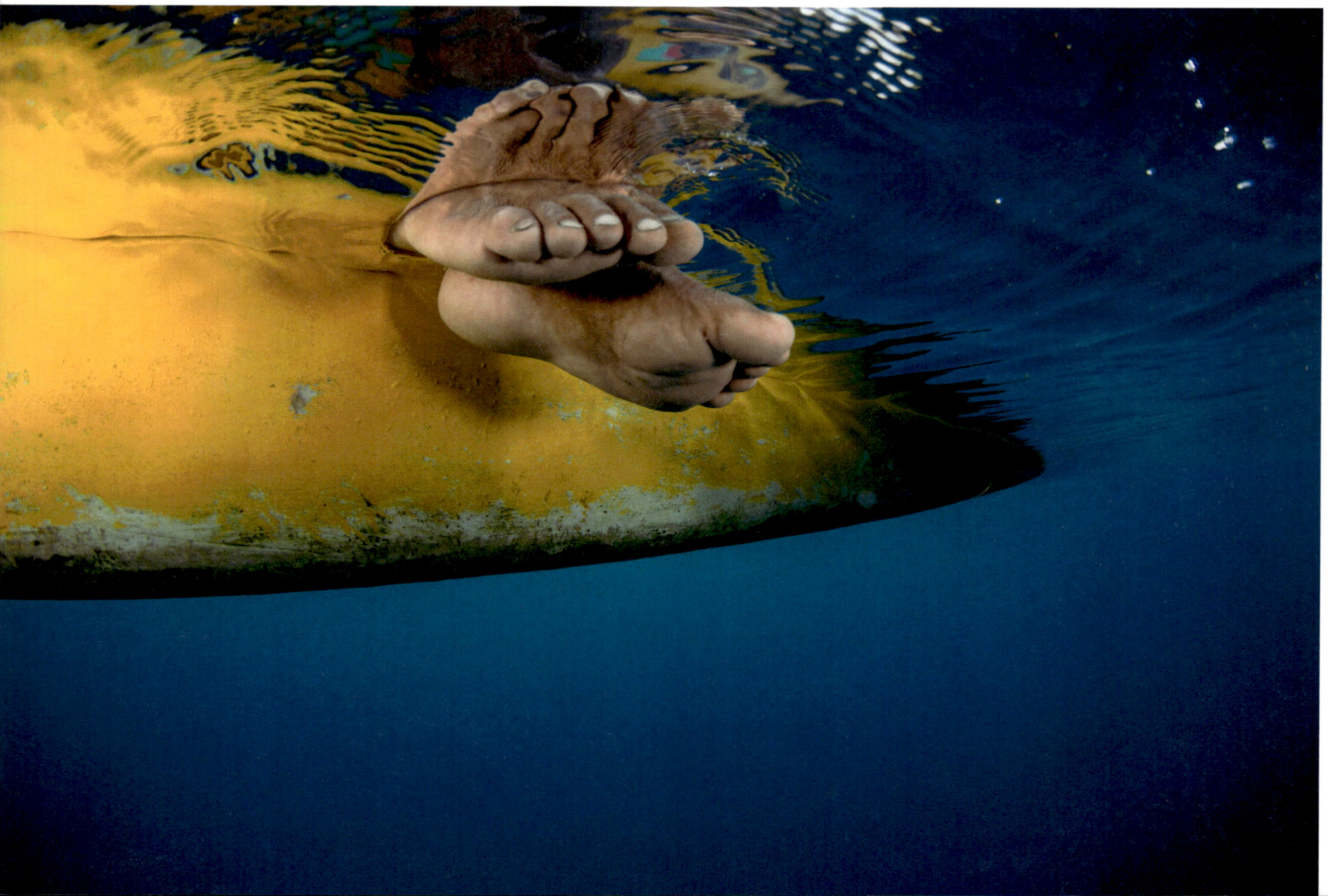

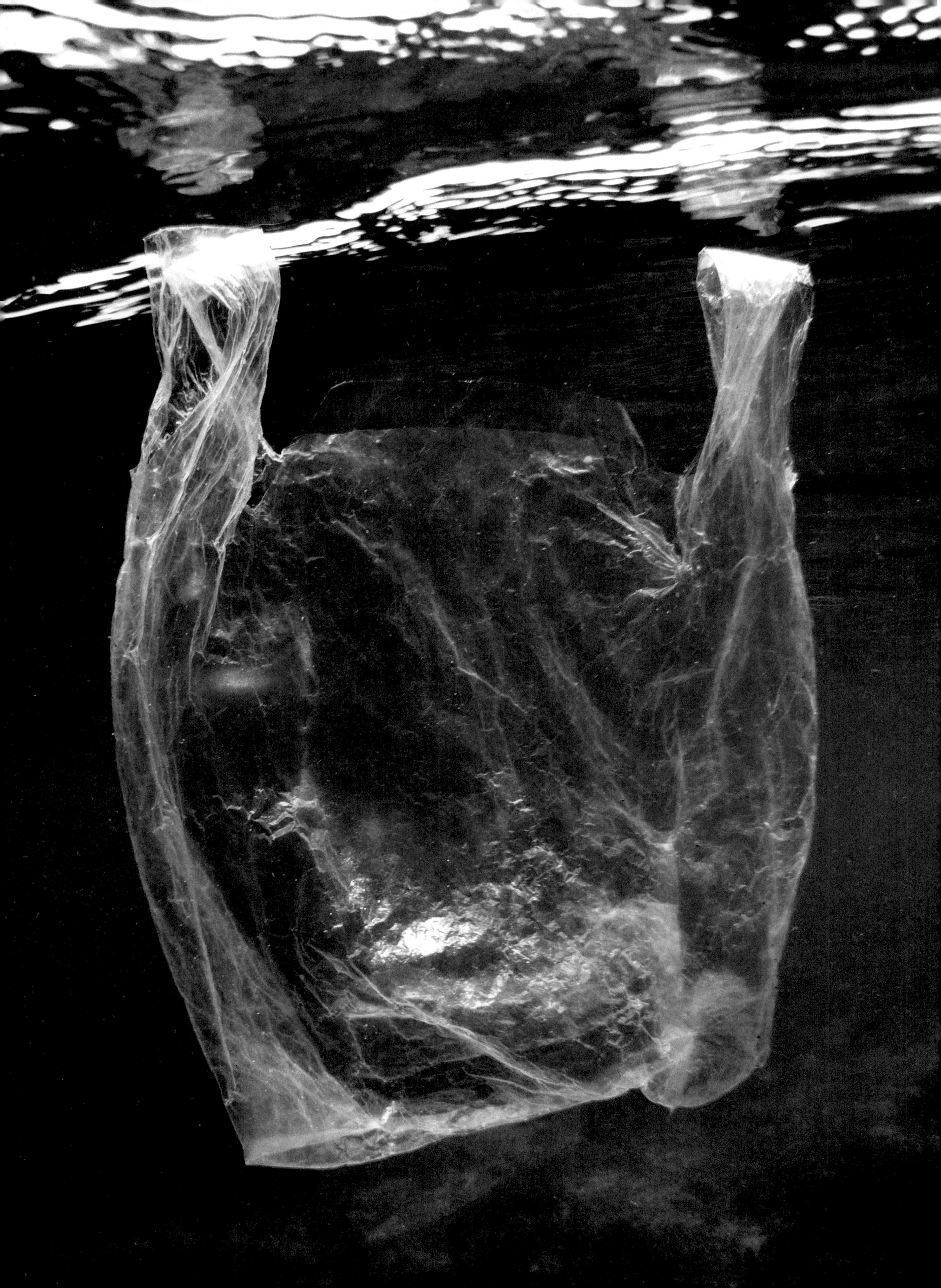

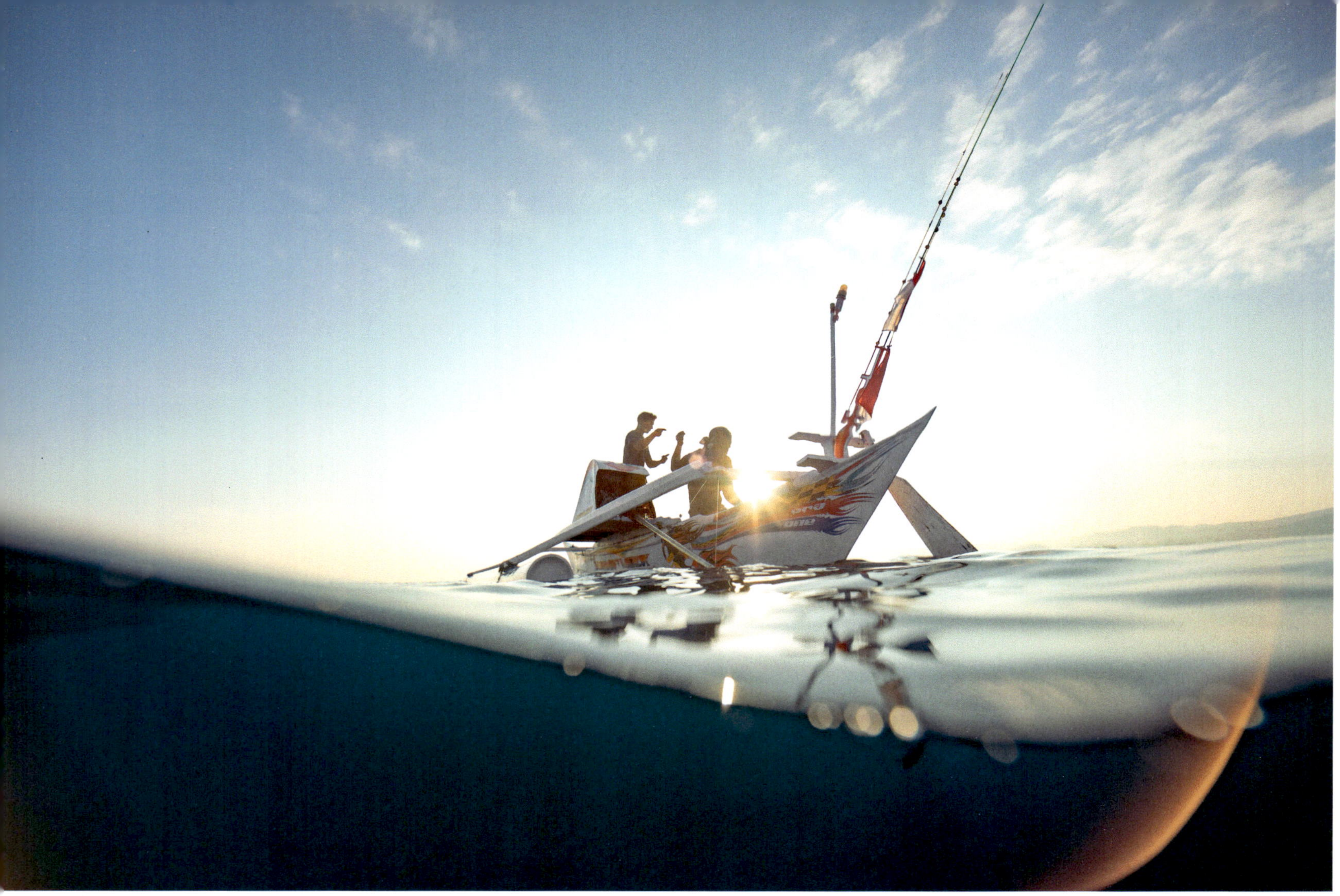

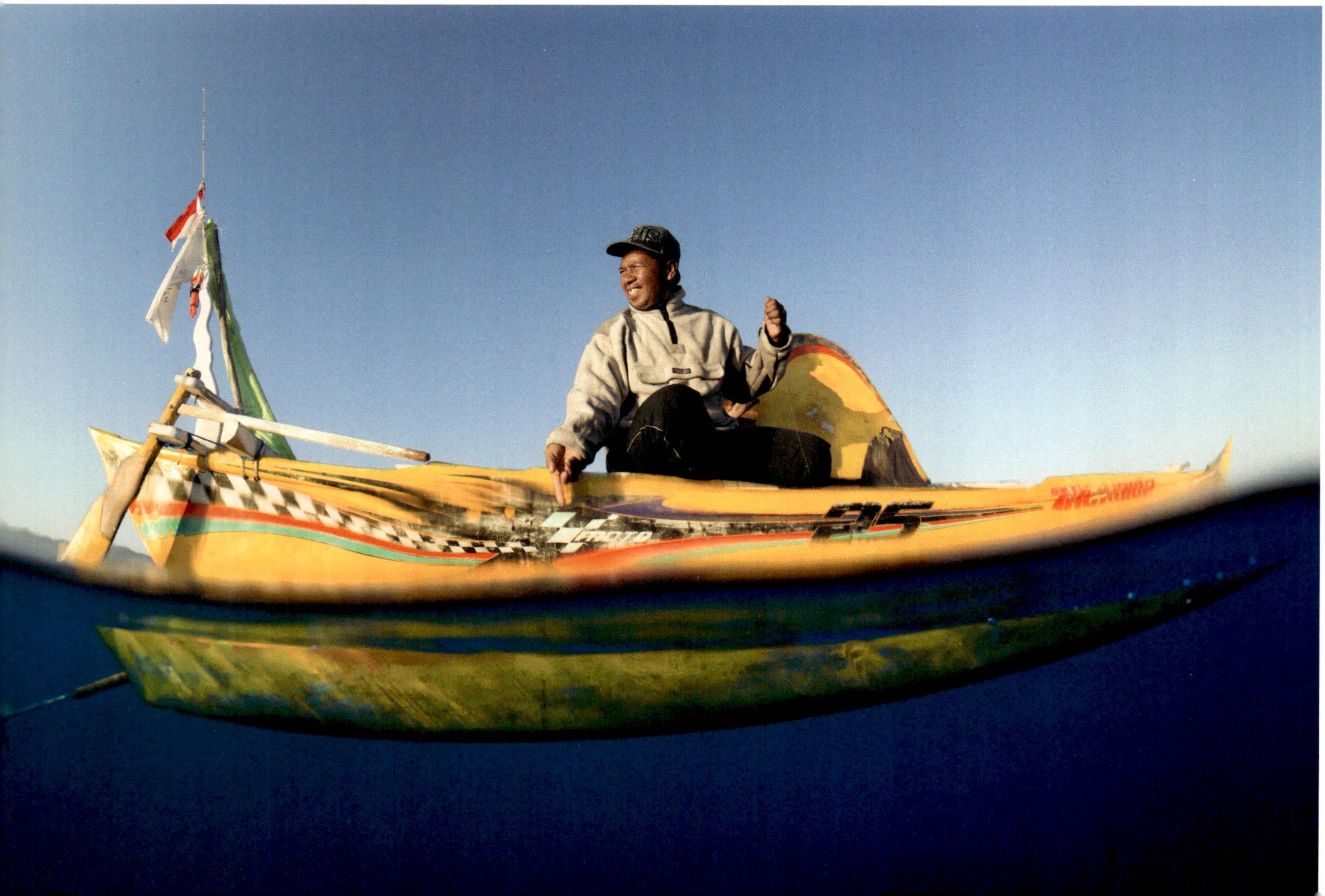

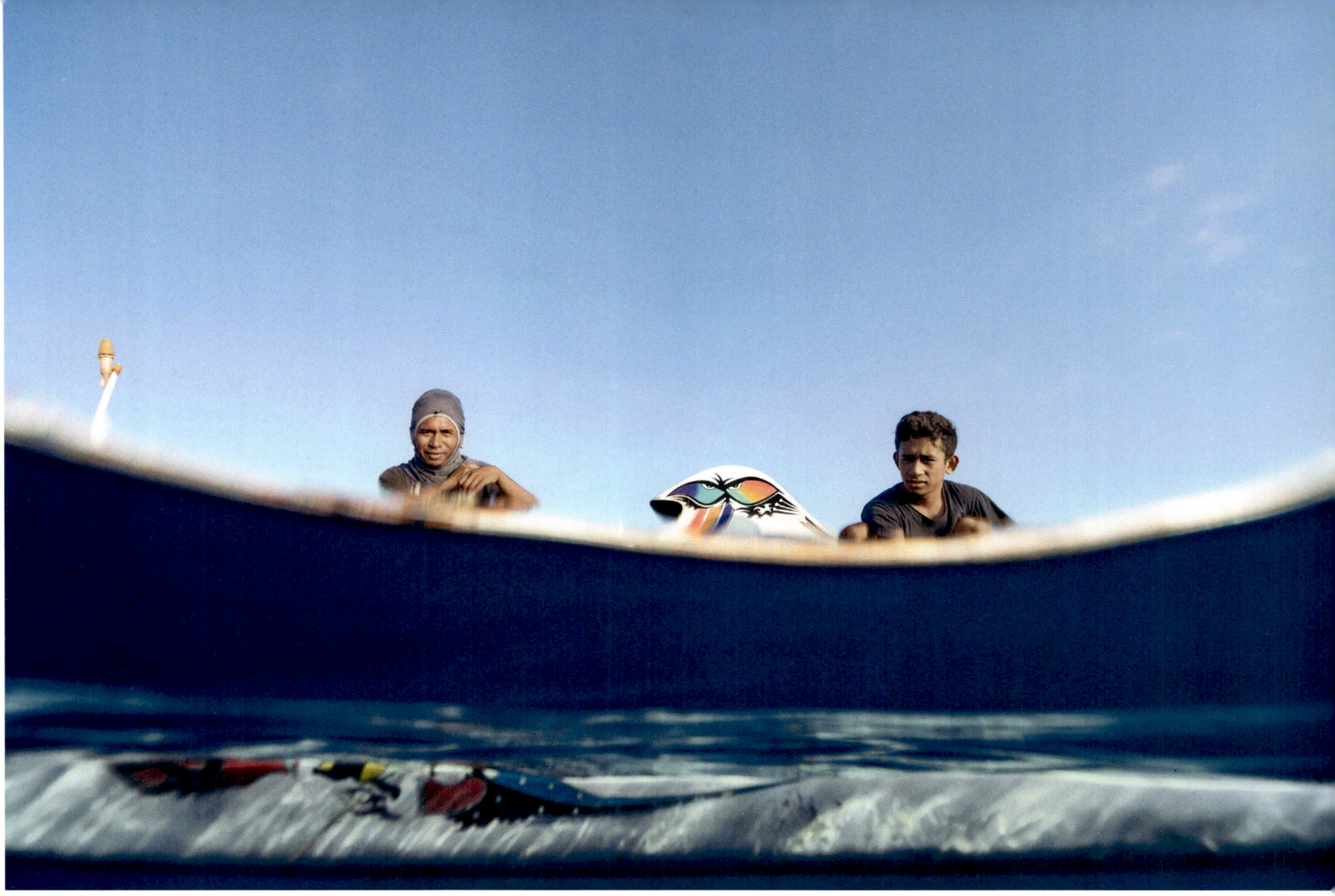

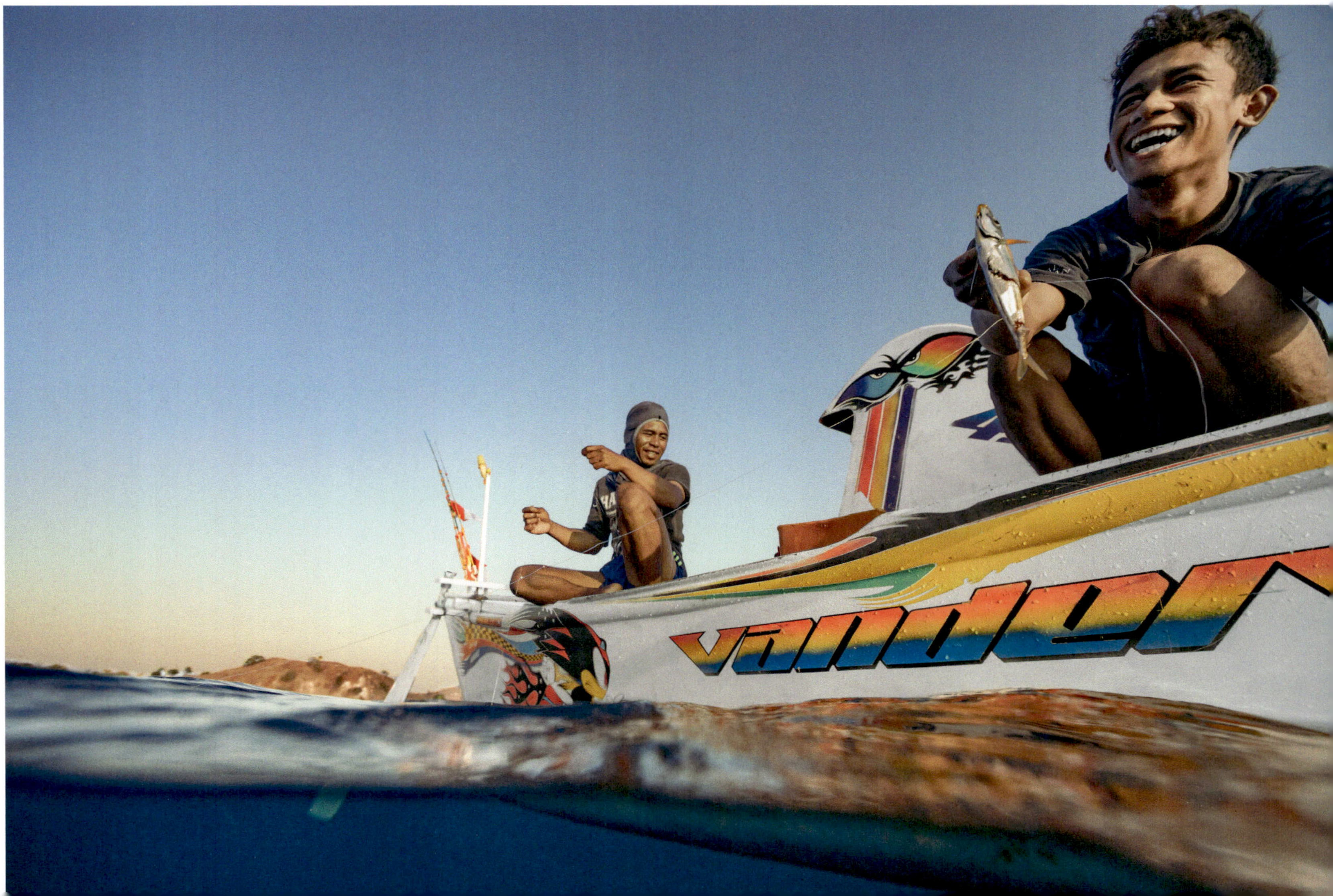

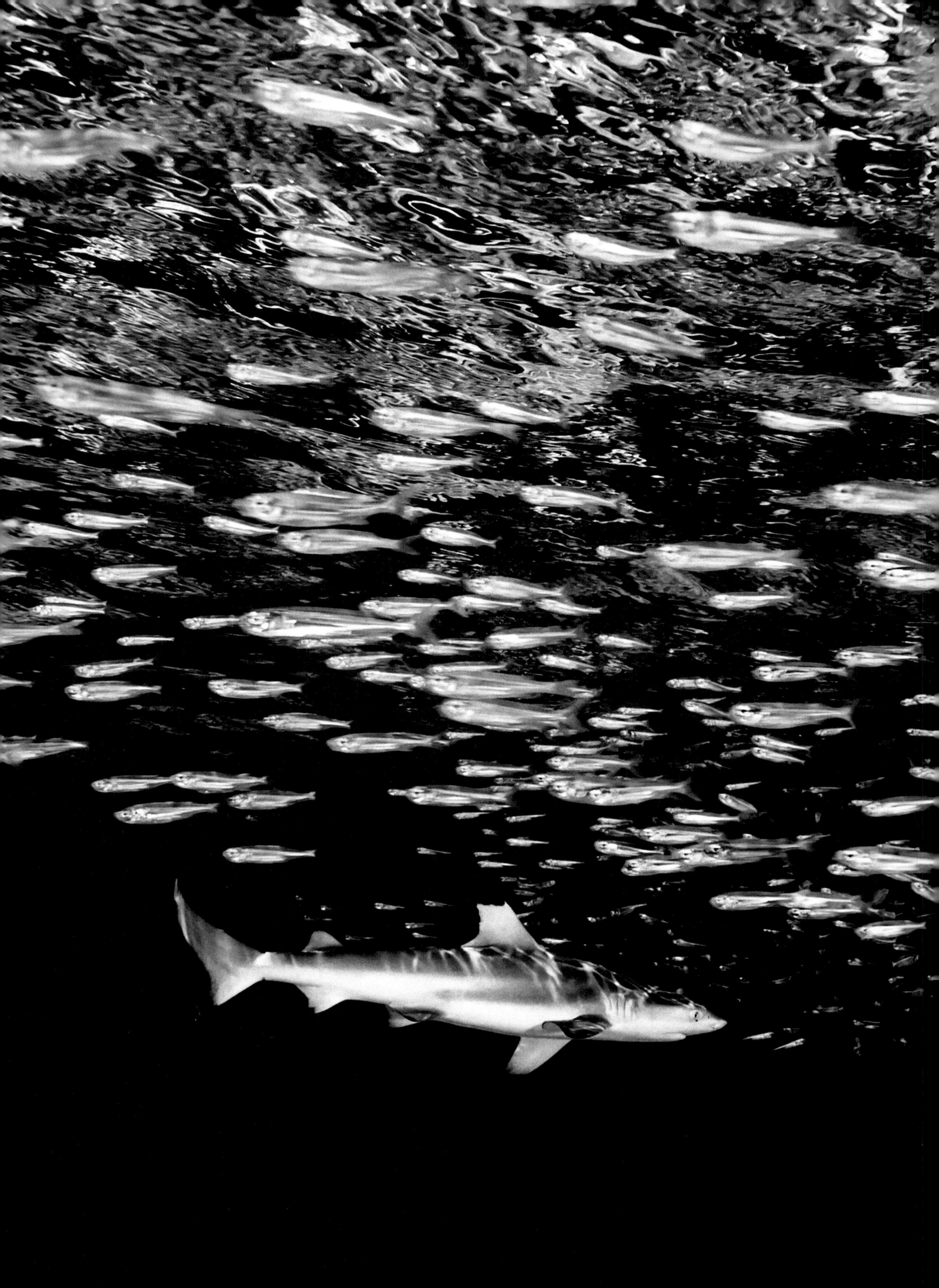

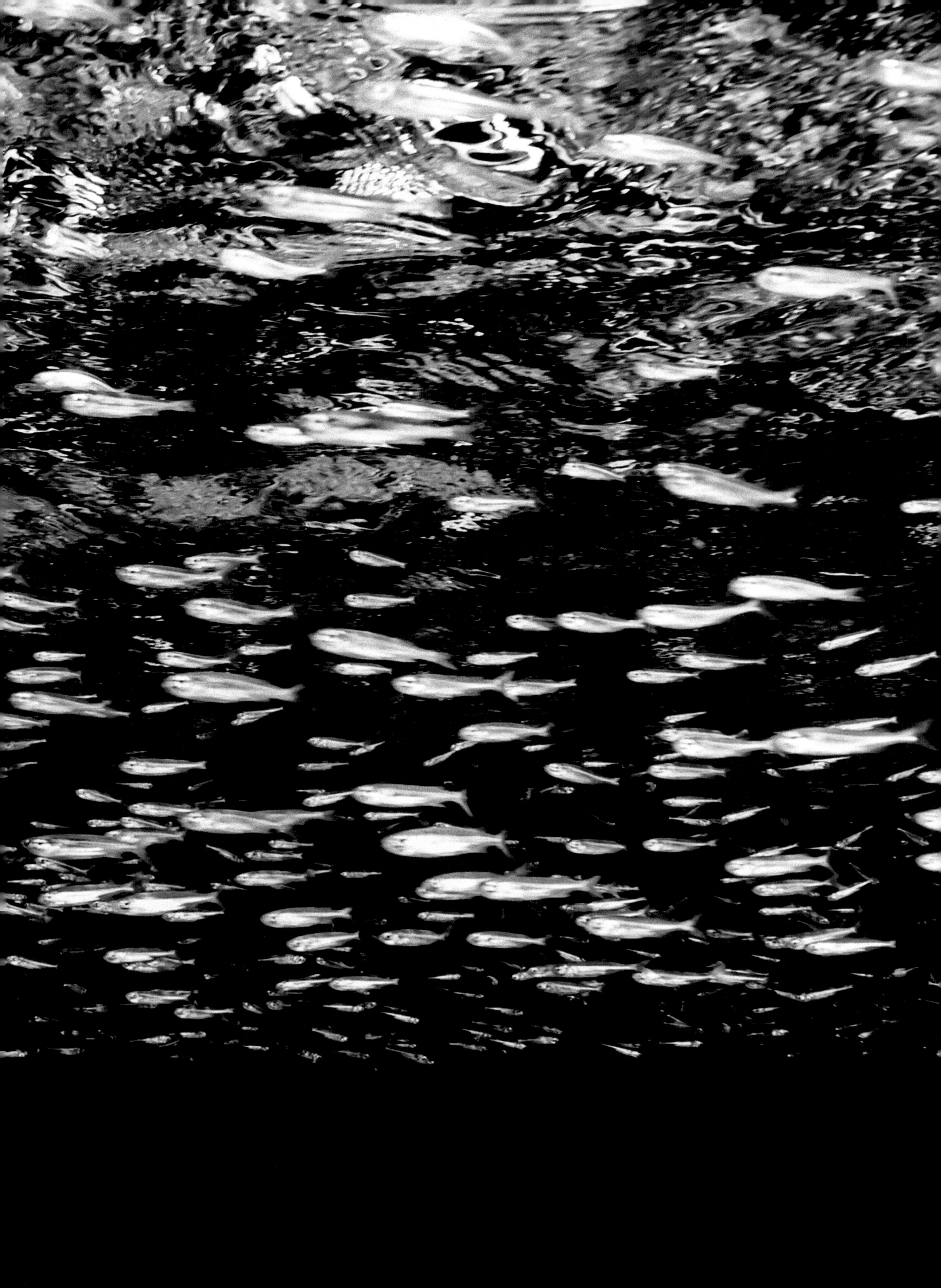

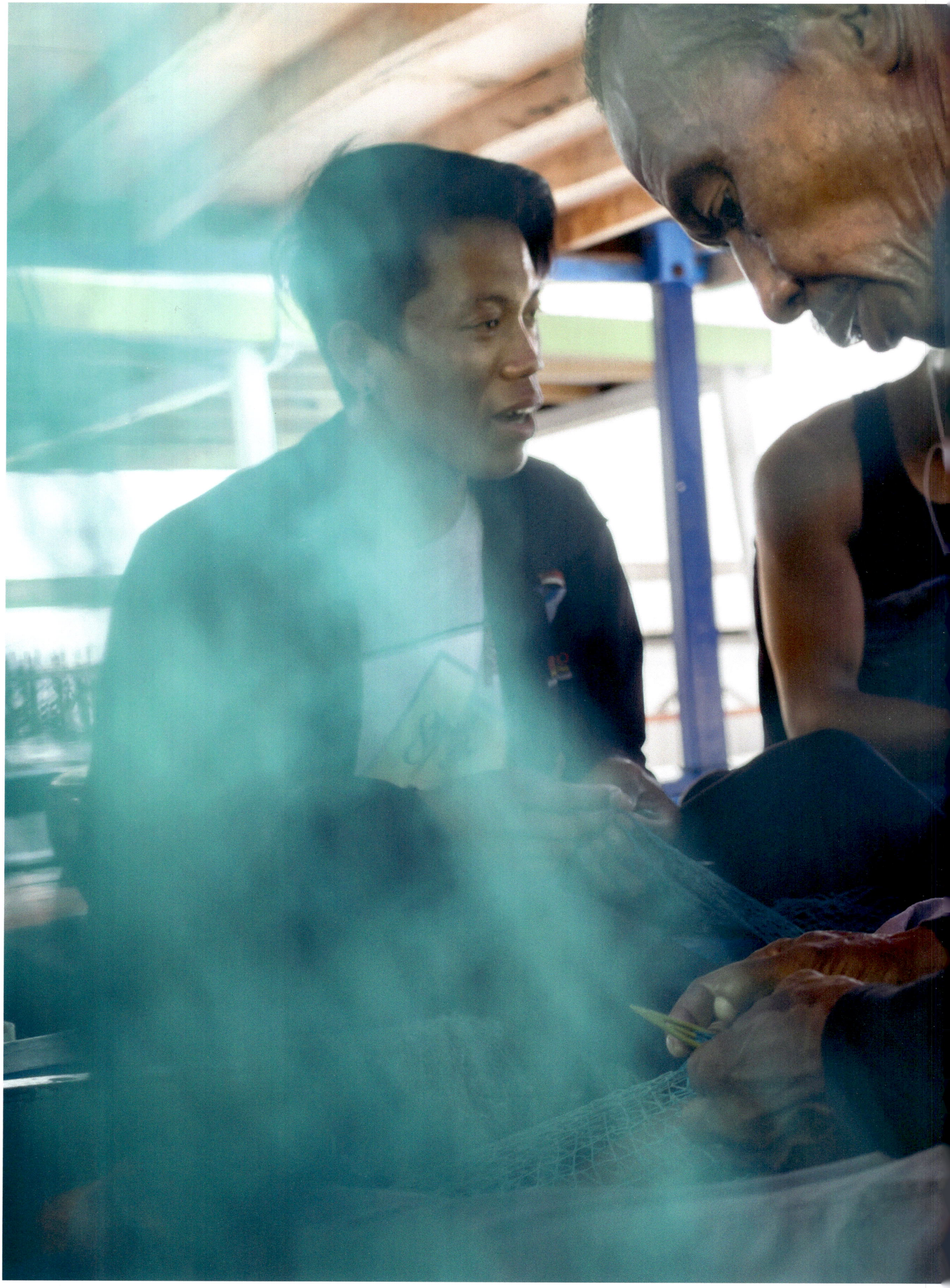

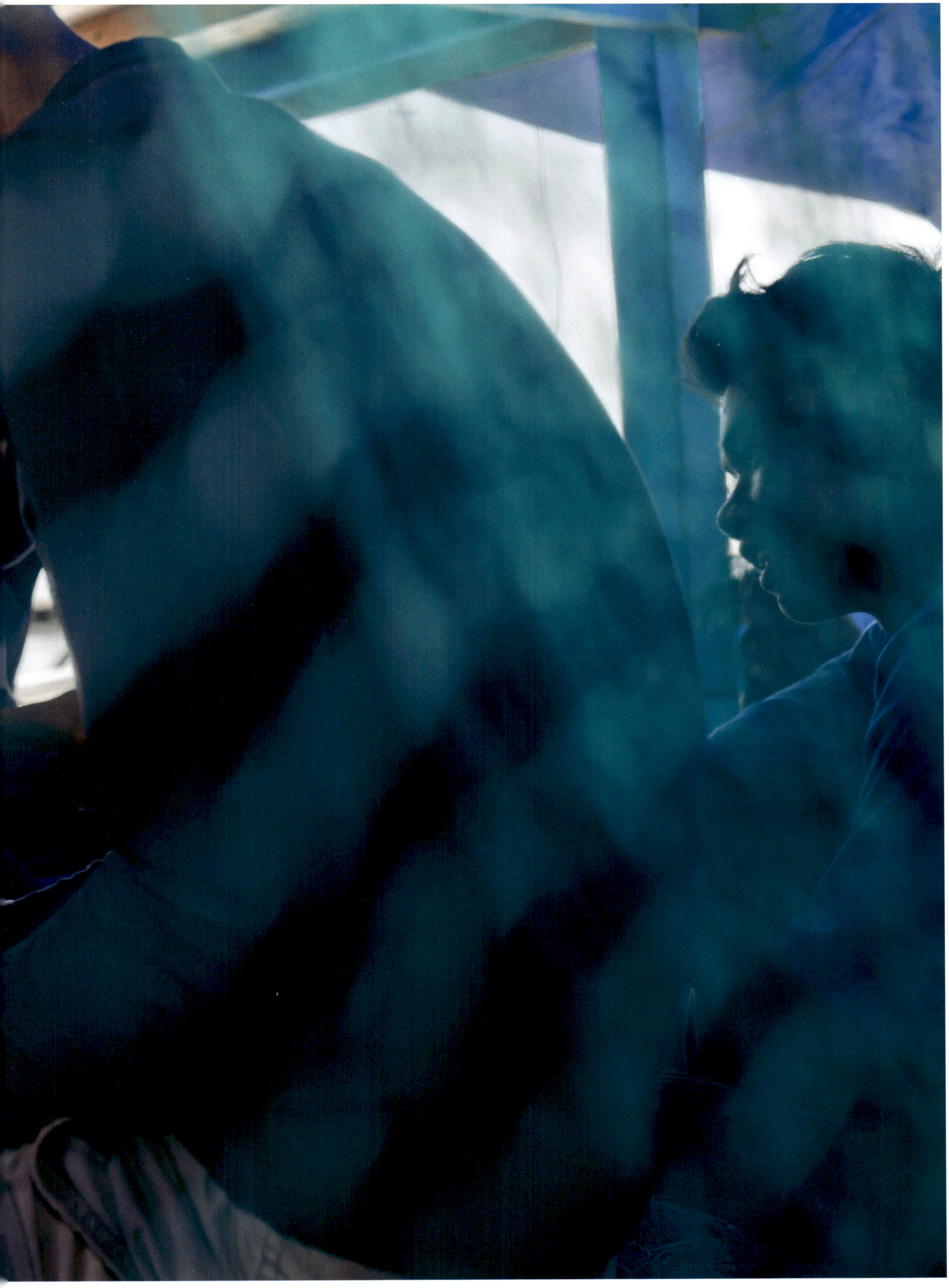

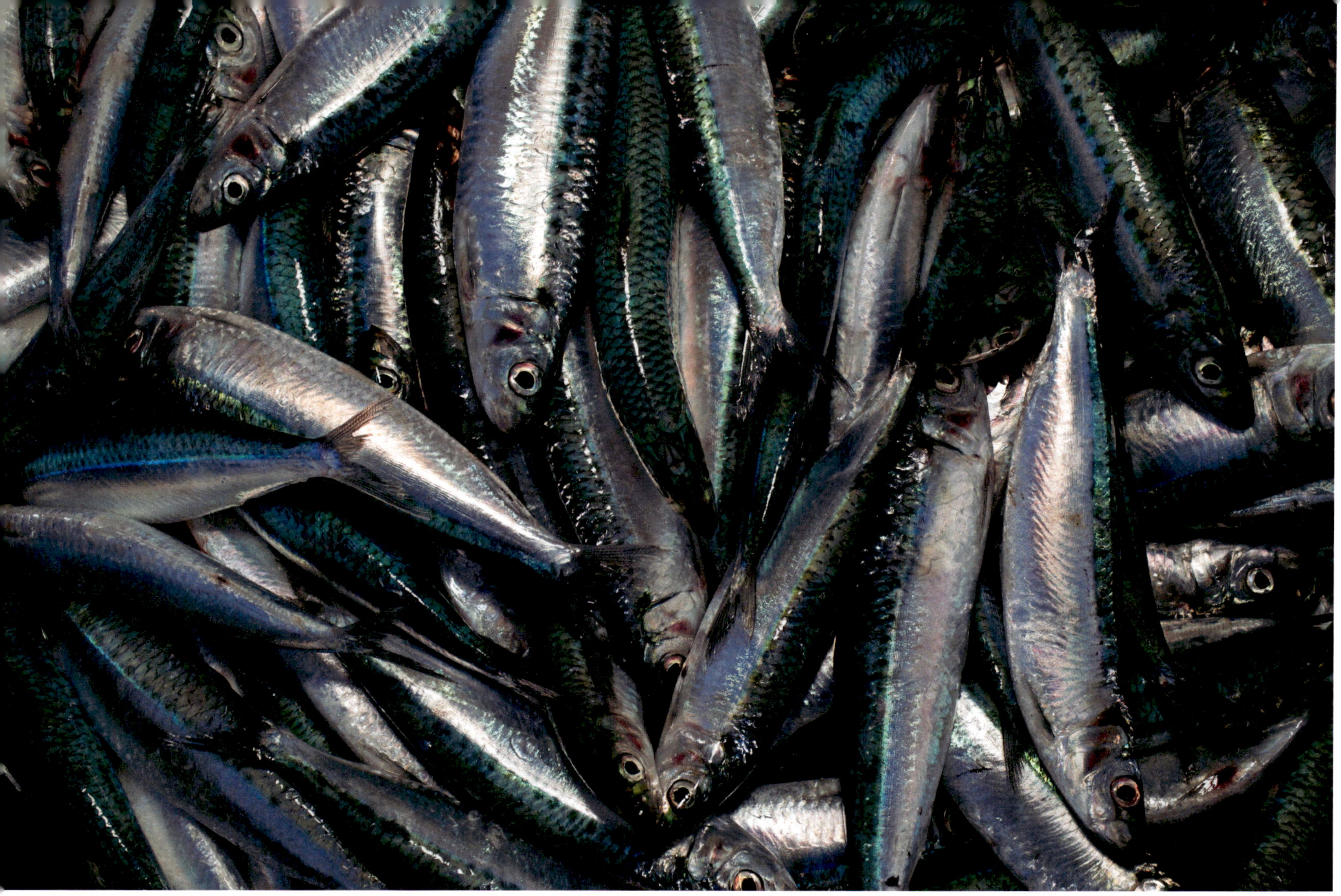

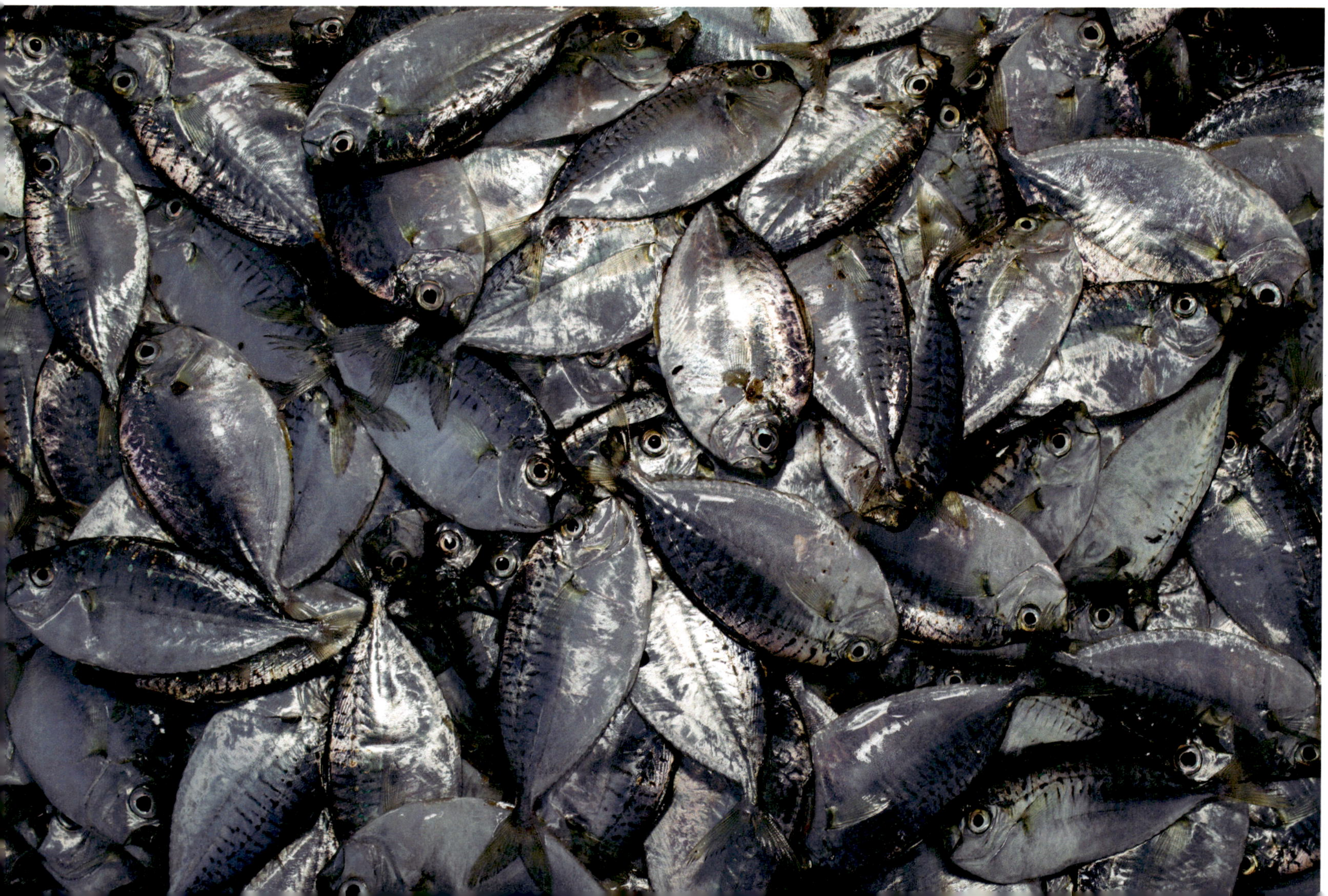

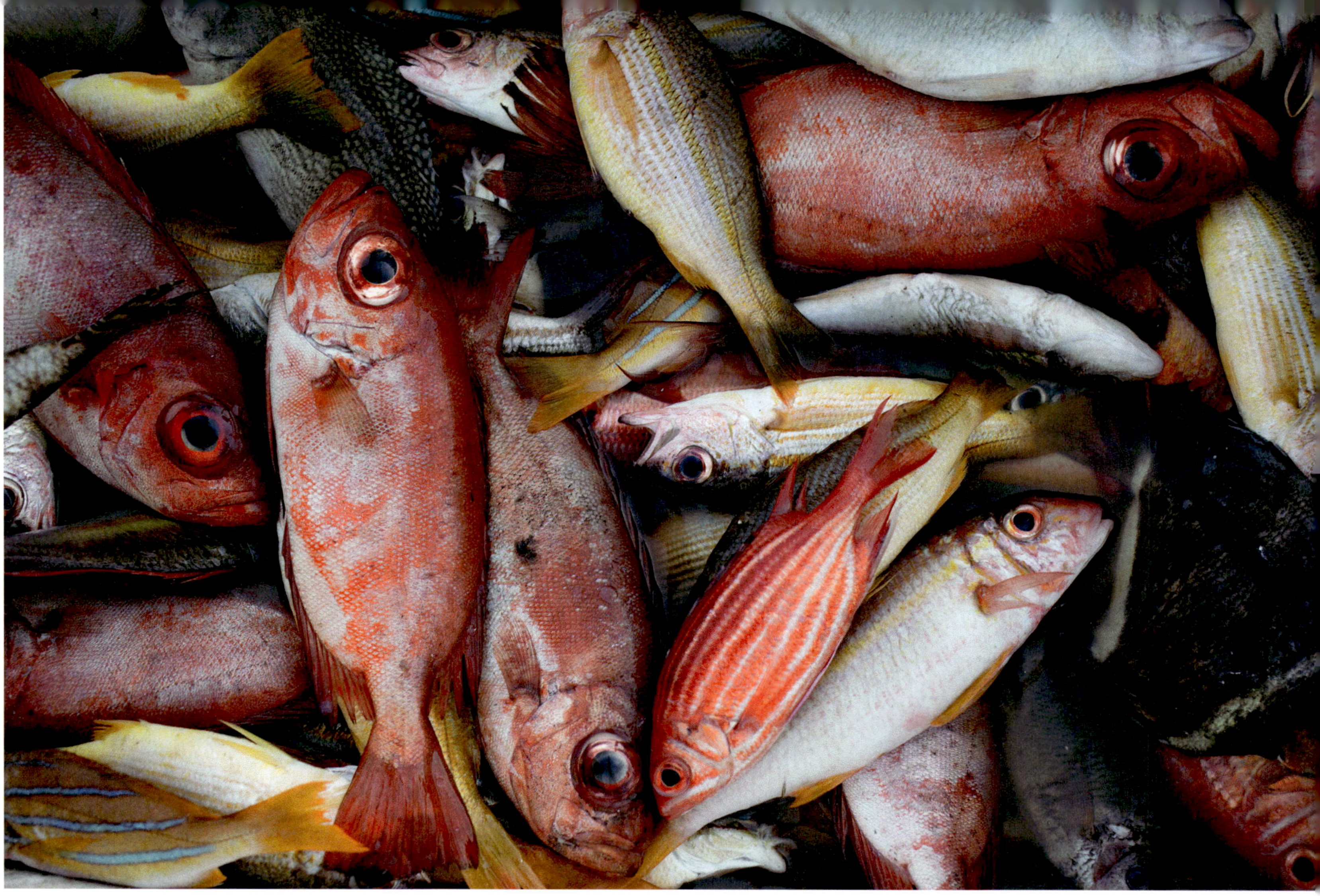

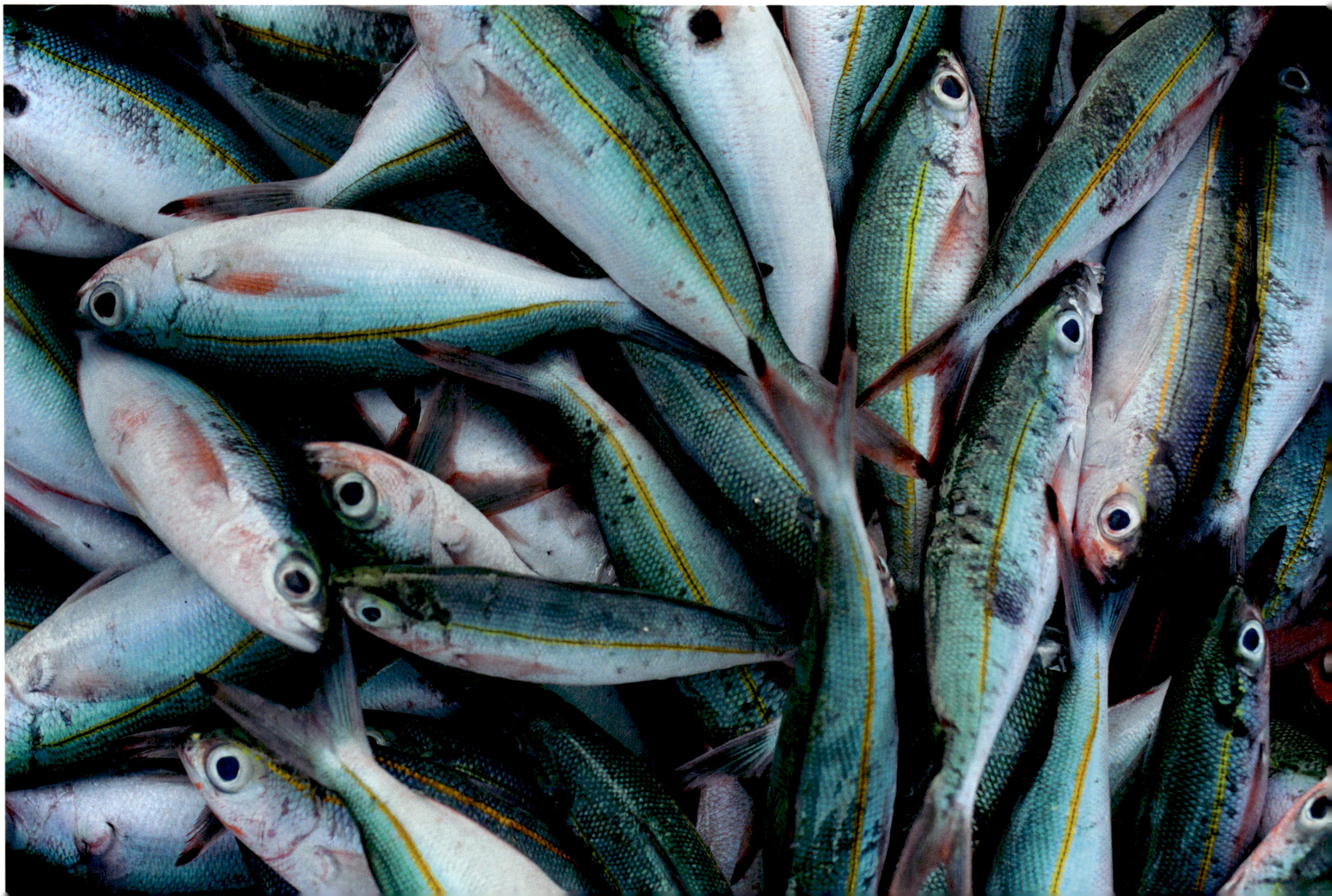

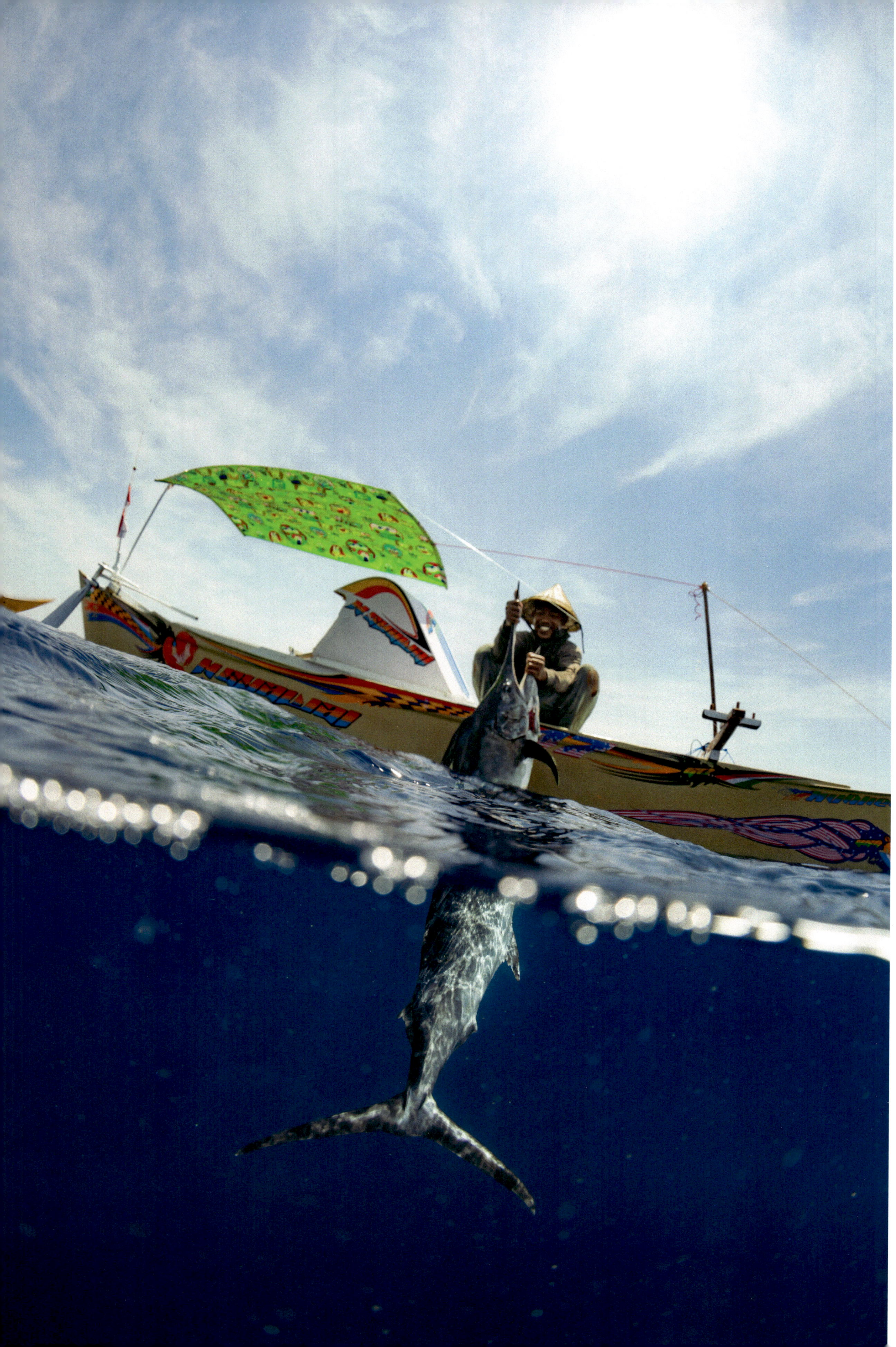

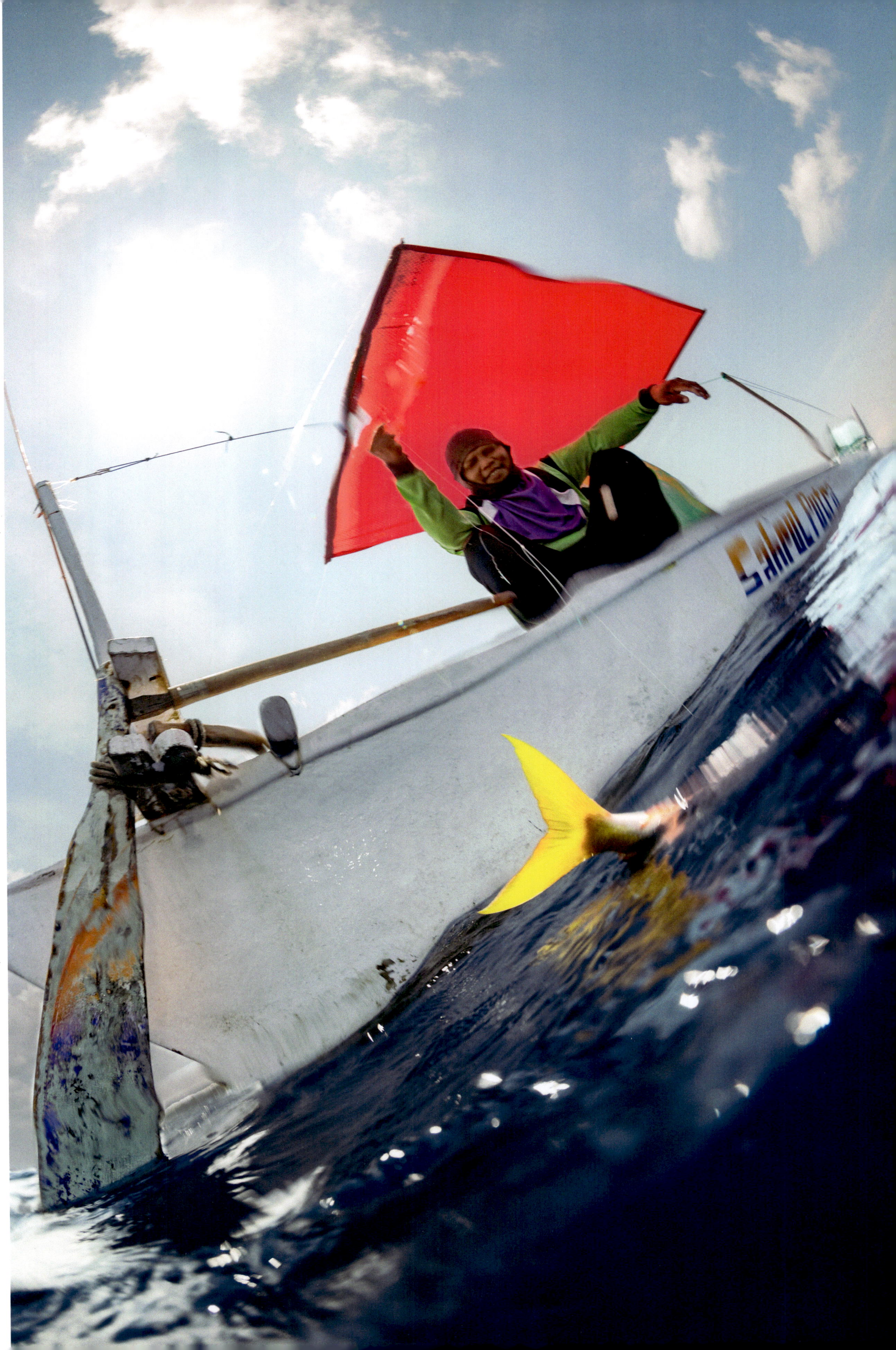

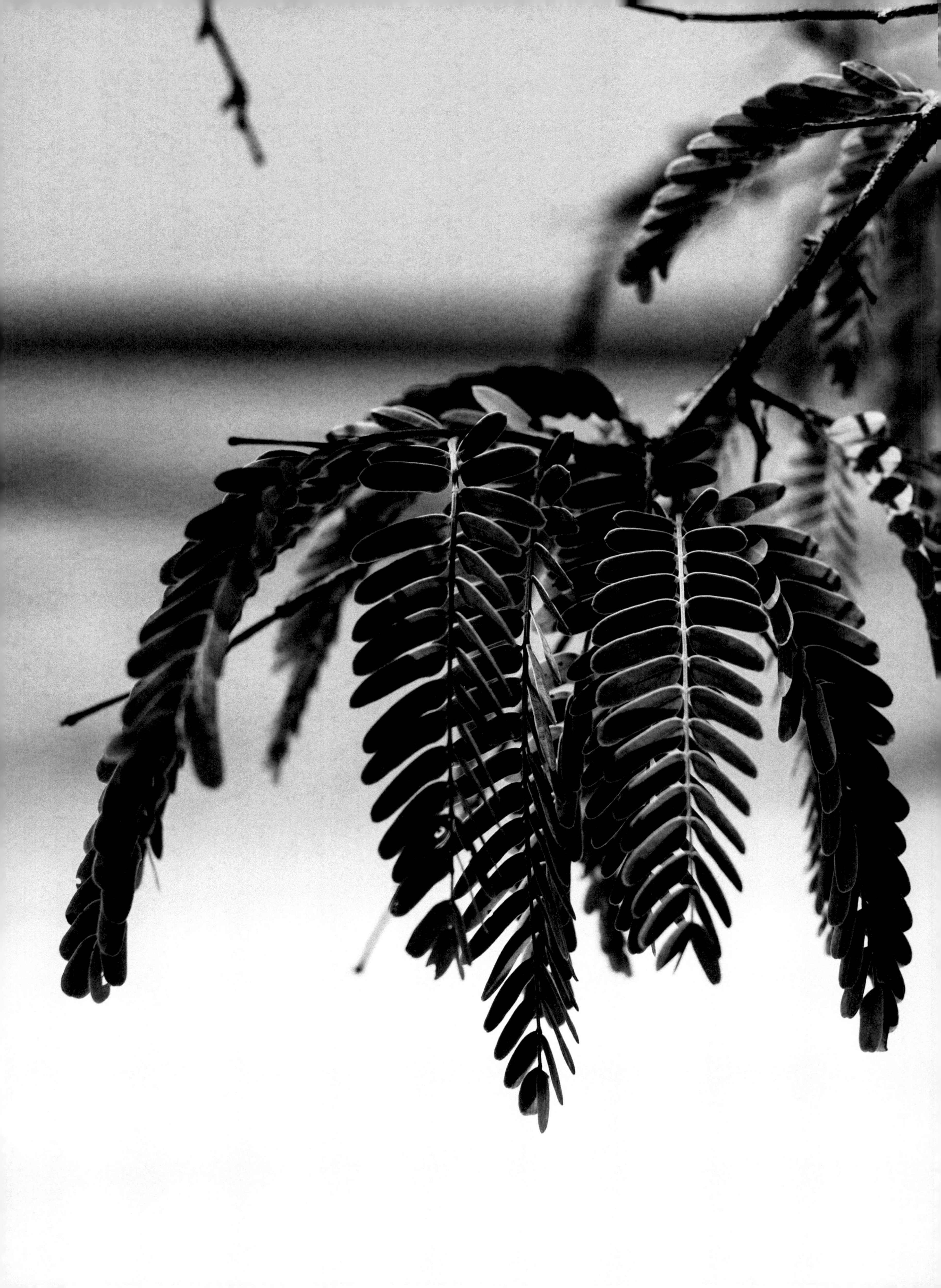

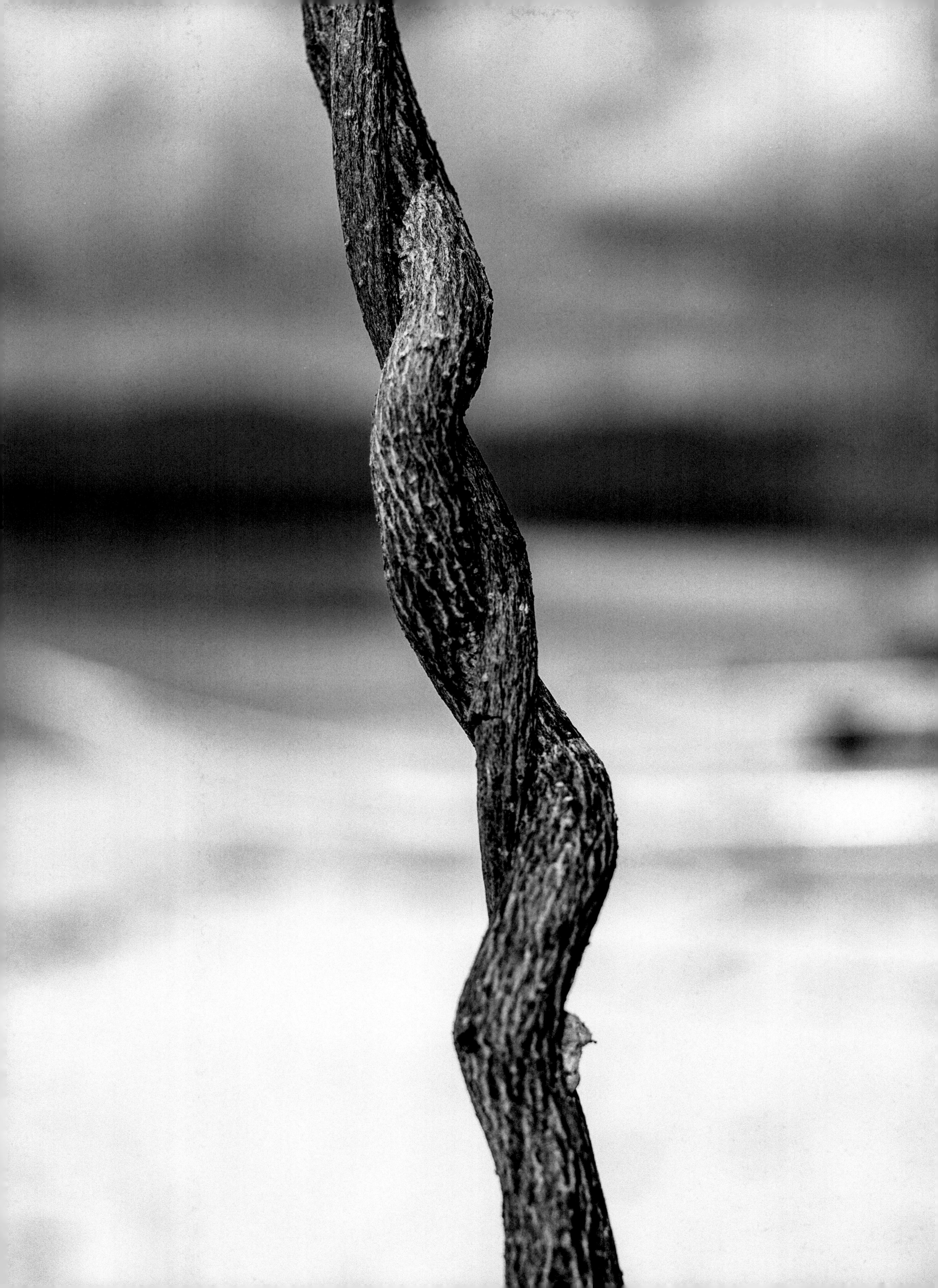

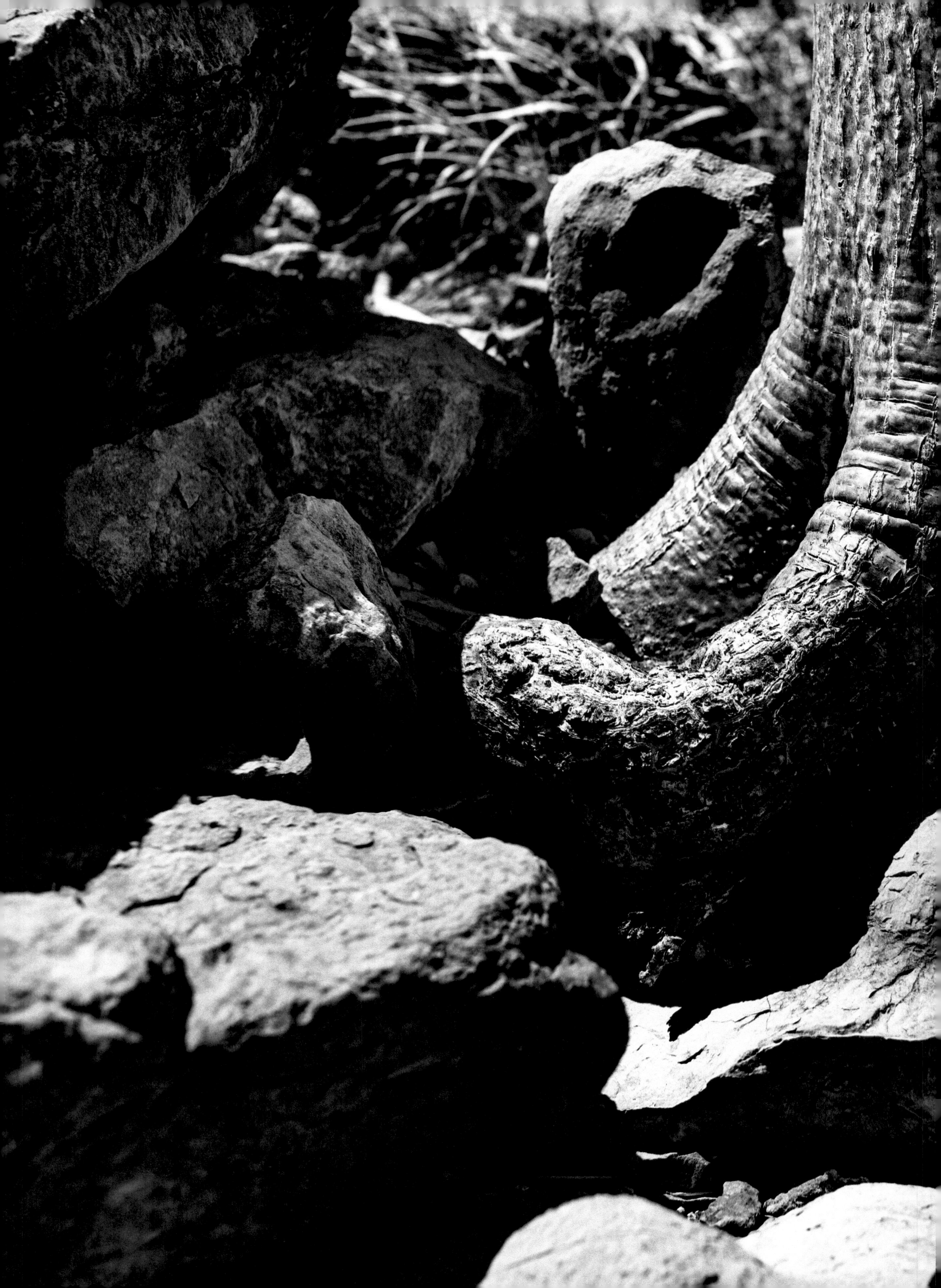

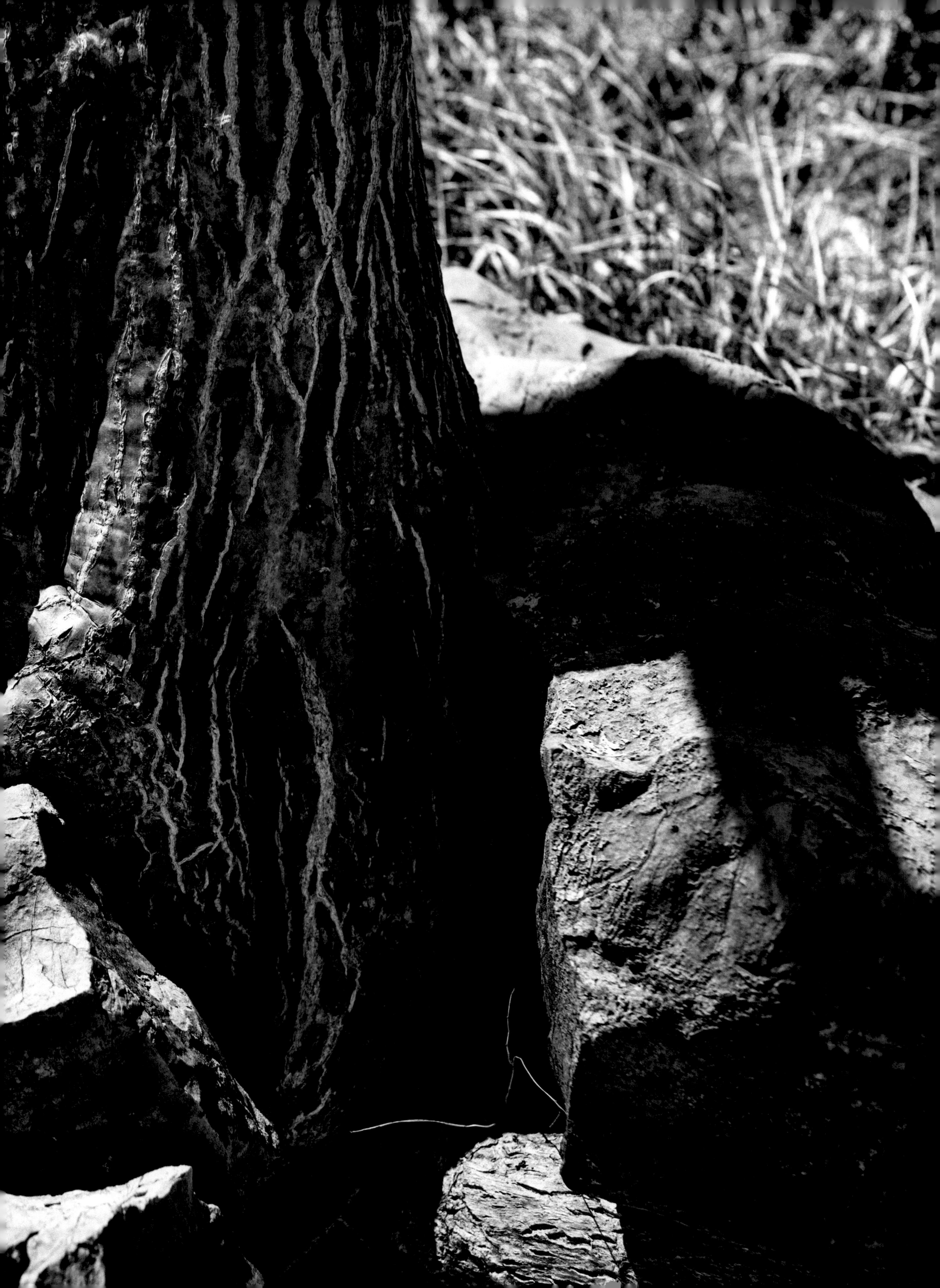

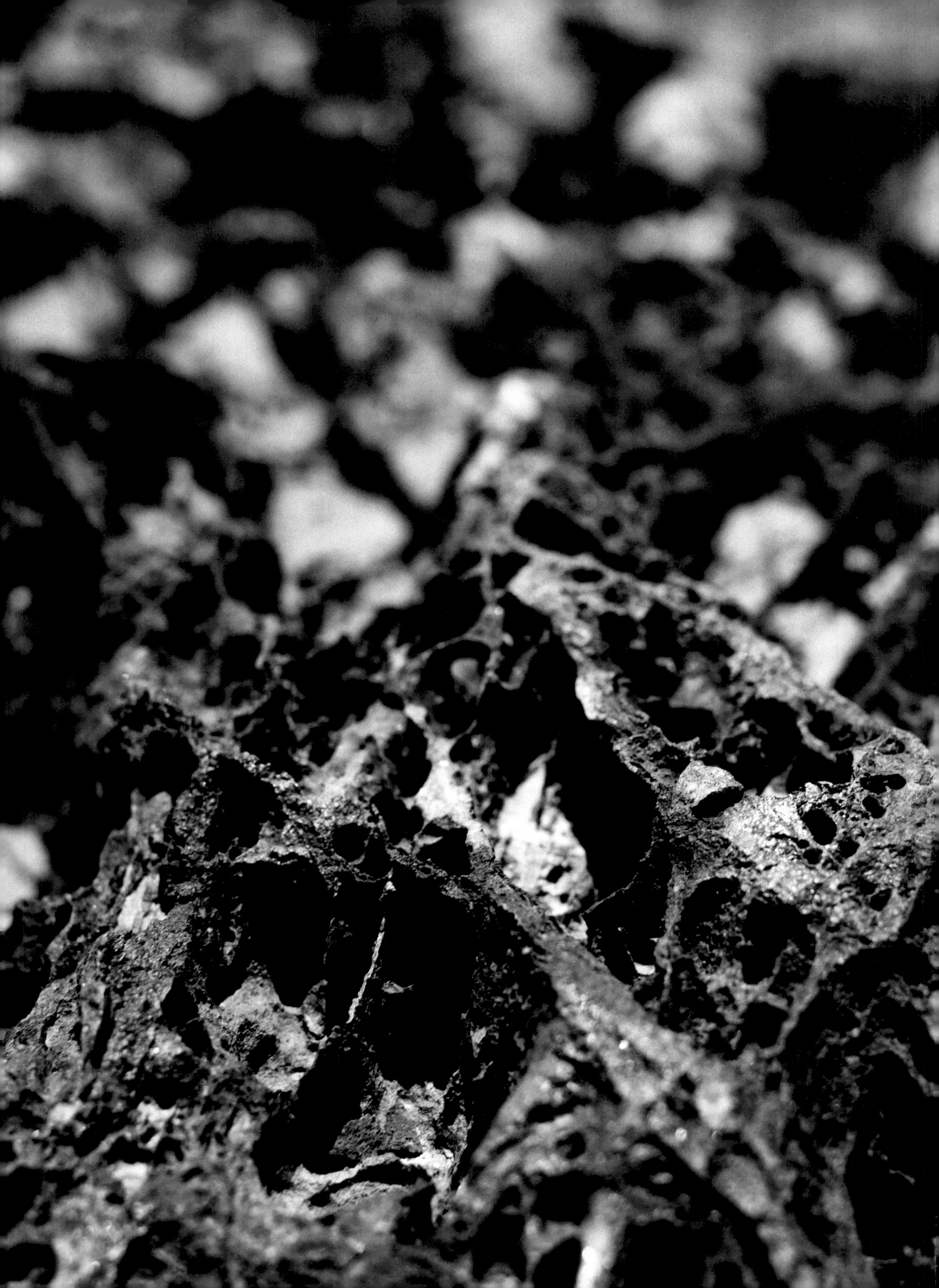

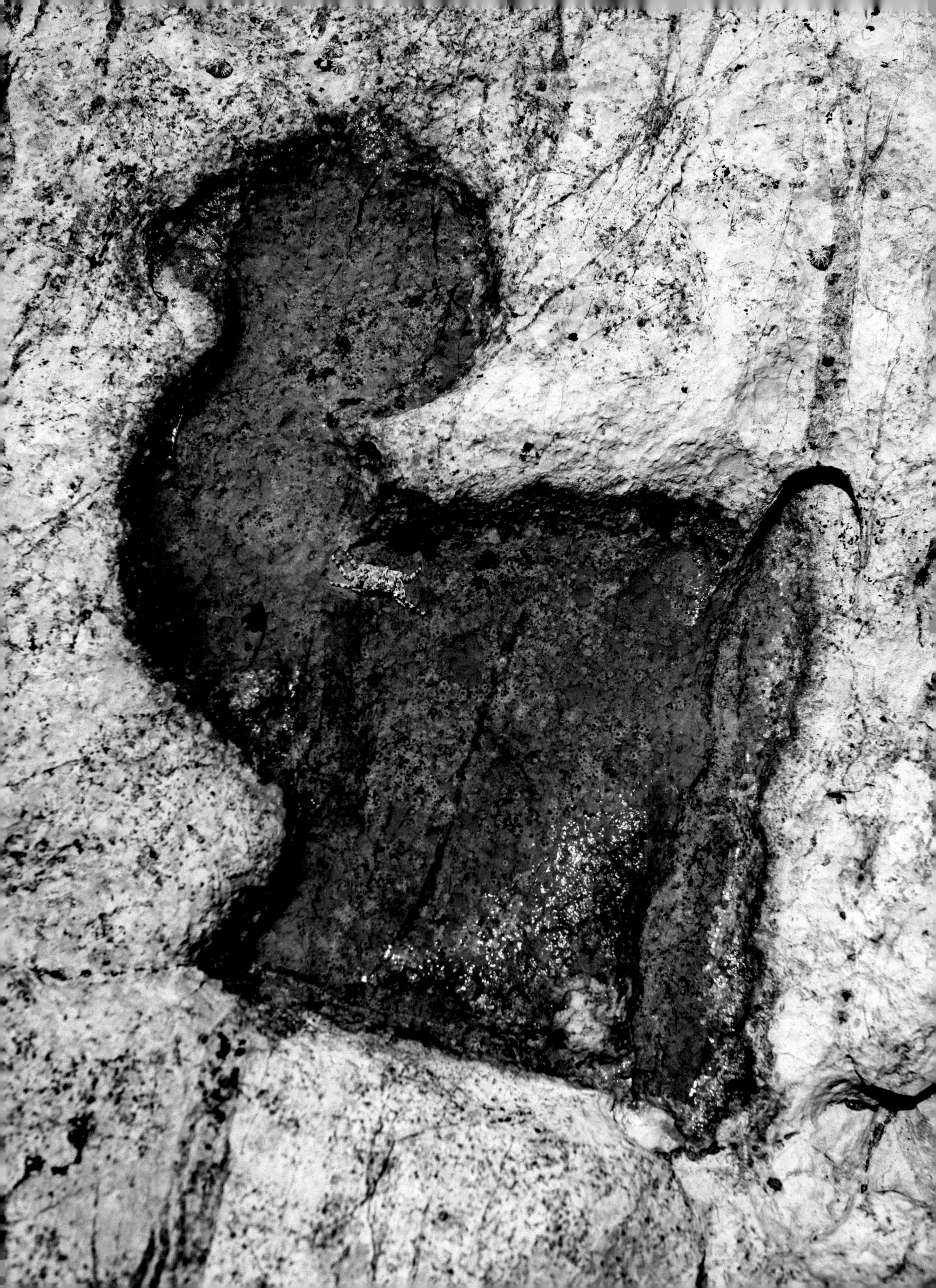

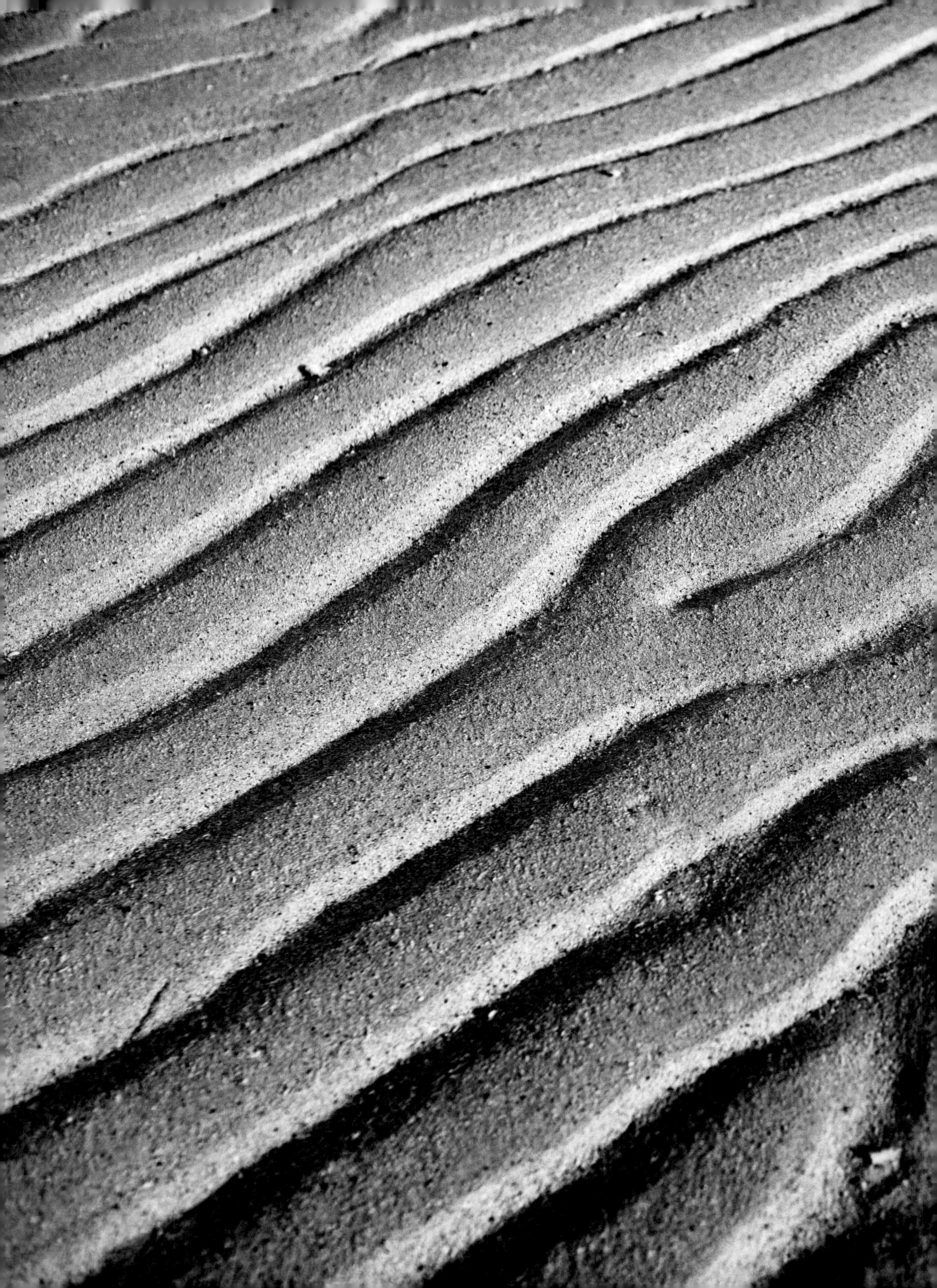

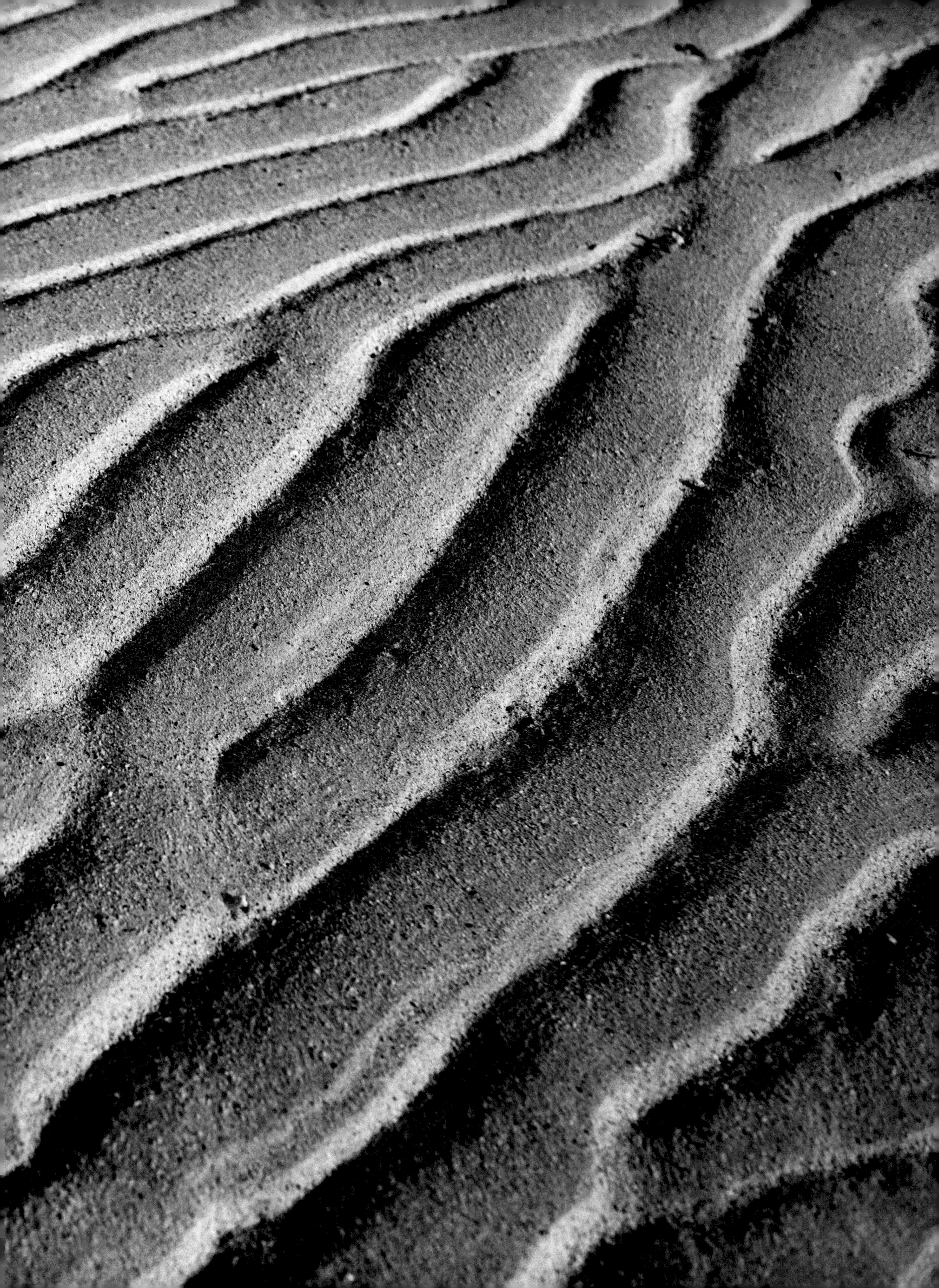

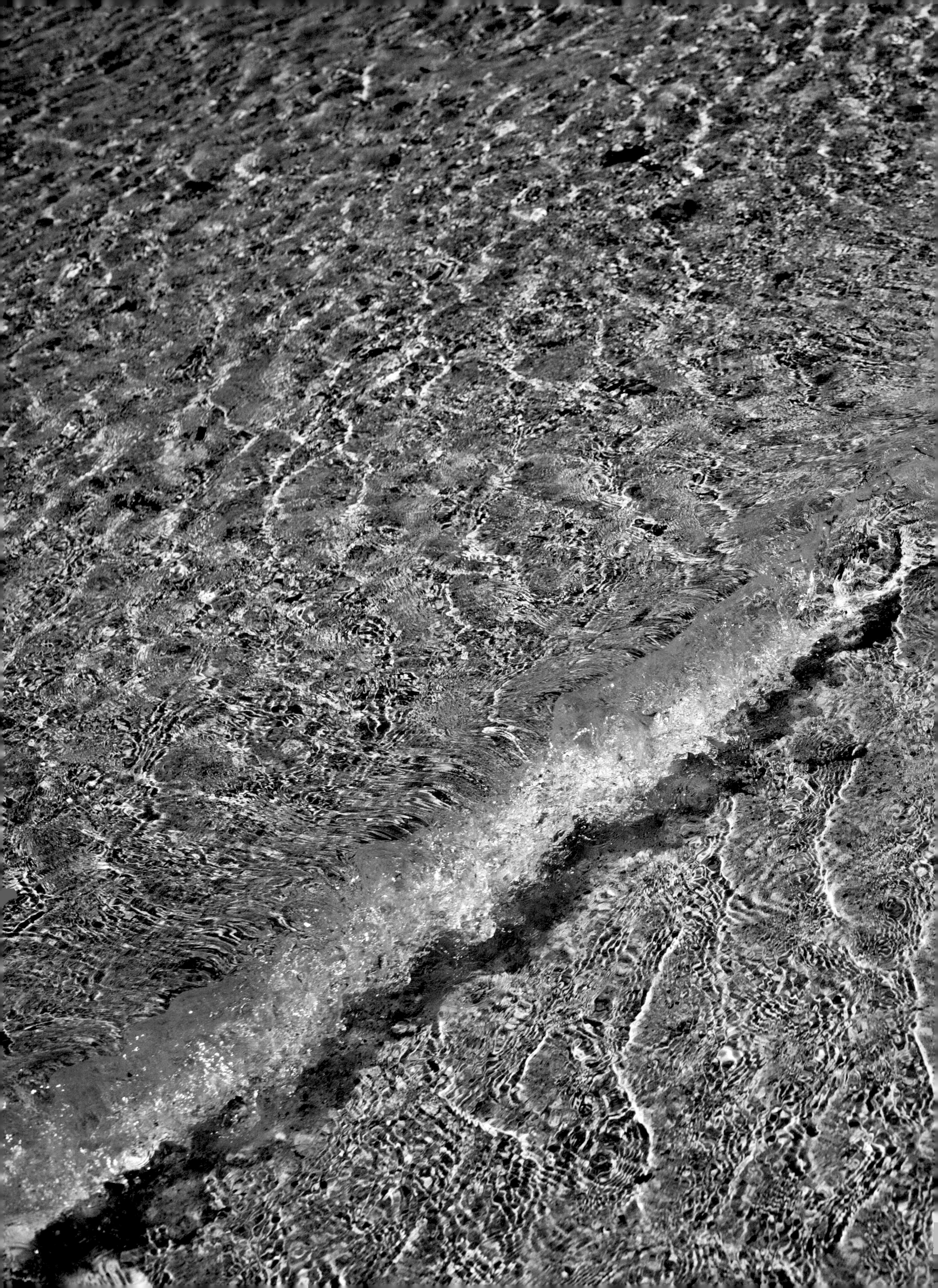

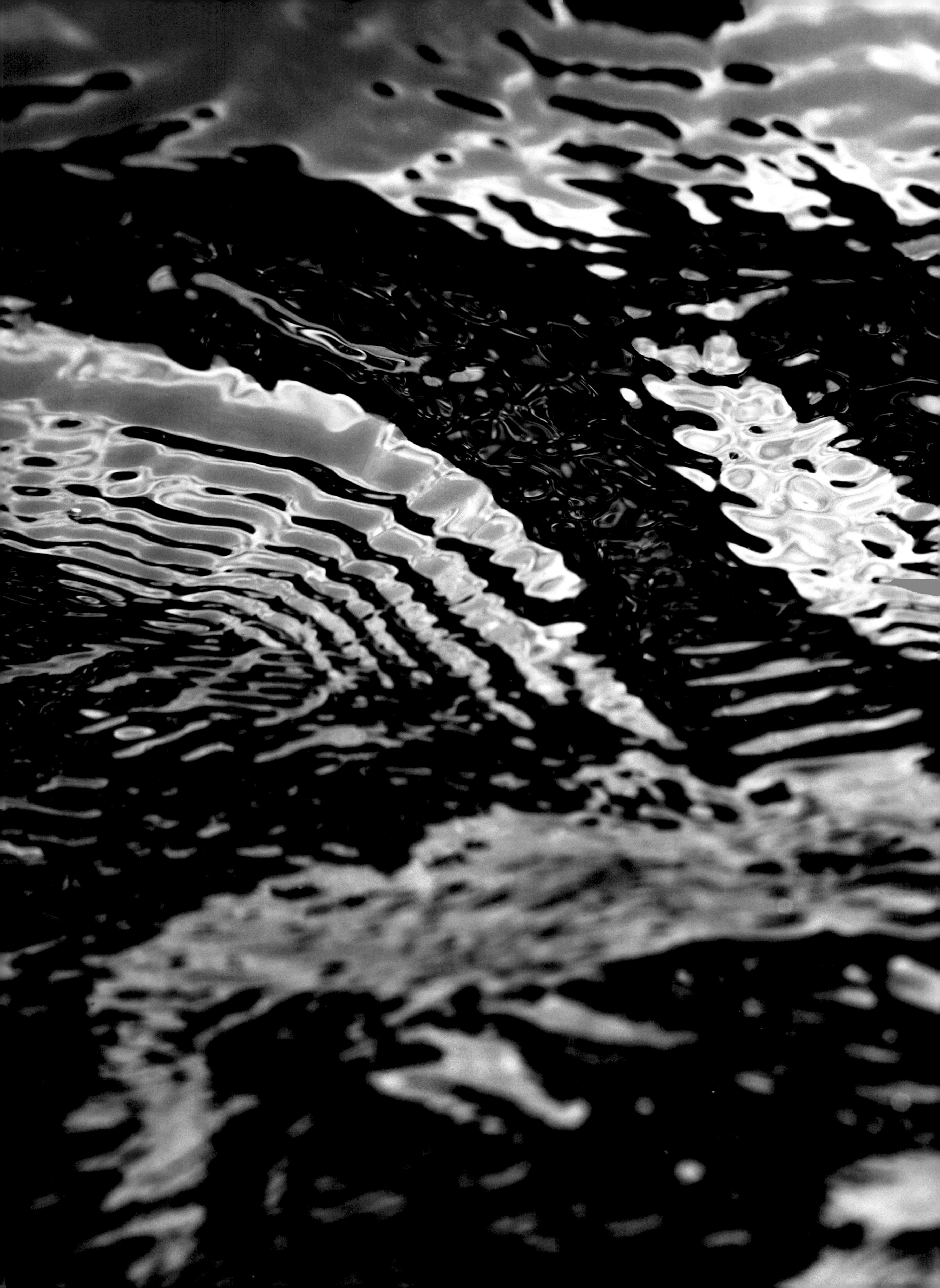

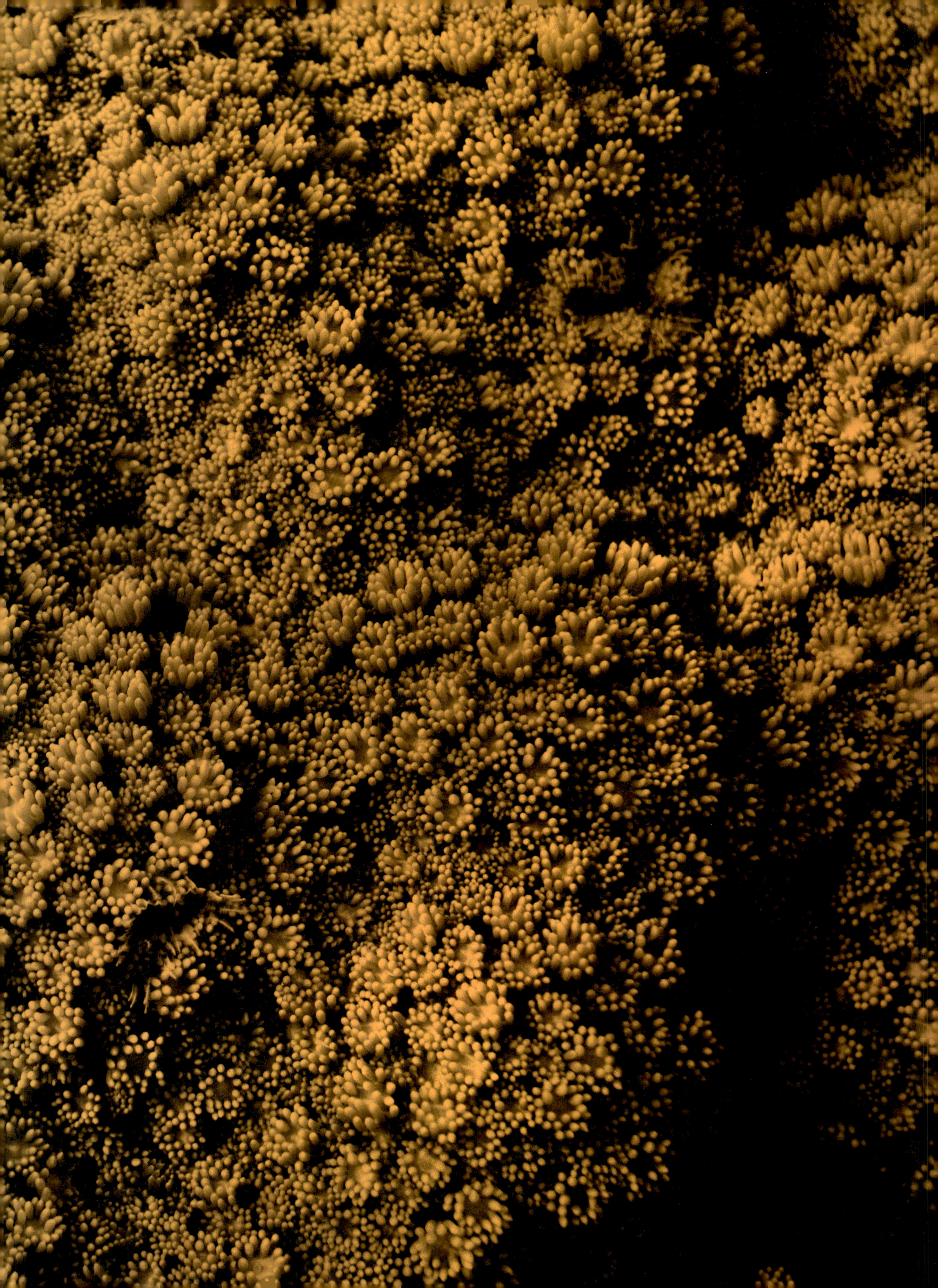

THE THREAD OF LIFE

Martin grasped this increasingly tenuous thread linking coral and humans. He grabbed it on a whim, a sudden realisation. After a childhood spent on the shores of the Mediterranean, he realised that he cherished marine life too much to be able to continue working in the coral and fish factory farms where he began his career. He fell in love with Indonesia, a country which welcomed him, and he refuses to be a part of the legal pillaging of the country's natural resources.

Protecting coral reefs by interacting with living things is now the thread that guides Martin. After years of research, experimentation and unexpected meetings, Coral Guardian was born. Three years after Martin co-founded the organisation in 2012, its Franco-Indonesian team discovered Hatamin Island, built a relationship with the people living in the fishing village of Seray Besar and launched their first participative conservation project.

The restoration project operates on a local scale, together with the residents of the village who live off the reefs. It begins by opening a dialogue with traditional fishermen who have to be convinced that protecting the reefs will help them maintain a sufficient level of fishing and ensure a better future for their families. It is built on sharing knowledge between the organisation and the fishermen, who later will lead training sessions themselves to teach what they know about coral aquaculture. The project's success can be measured by the total surface area of restored corals: 500 cubic metres; the return of biodiversity, with four to five times as many marine species; and an impressive increase of more than 1000% in fish stocks.

Martin lived with the village's inhabitants each year for several months at a time, over the course of six years. He realised that he wanted to share his experience, and his artistic eye followed suit, his lens capturing images on the fly. His photographs mix what is imaginary with what is real, beliefs and innovation. They plunge us into emotions, underwater and on land. Martin tells his story, recounts the riches of his experiences with the fishermen, recounts their friendship and their shared joys. We learn that it is they who are the heroes of this story. The ones who will keep the thread of life going for future generations.

Why should we protect coral reefs?

THE BASIS OF MARINE BIODIVERSITY
Scientists estimate that coral reefs are home to more than 25% of marine species. Other ecosystems are also created from coral.

COASTAL PROTECTION
Reefs make up only 0.2% of oceans. However, they protect more than 150,000 kilometres of coastline in more than 100 countries and territories. They form a barrier that absorbs the force of the waves and therefore help to reduce coastal erosion.

A FOOD SOURCE
More than 850 million people live less than 100 kilometres from a coral reef and benefit from the role they play in ecosystems. Around 330 million people depend directly on coral reefs. Reefs can produce between five and 15 tonnes of fish and shellfish per square kilometre.

A TOURIST ATTRACTION
Reefs are often a crucial part of the local economy in tropical regions. They attract scuba divers, free divers, recreational fishing and beach lovers.

A SOURCE OF ECONOMIC VALUE
Millions of people around the world depend on coral reefs to make a living. According to one estimation, the total annual profit made from coral reefs across the world is 29.8 million dollars.

A BOON FOR SCIENCE
Coral reefs also play a role in biomedical research, offering up different avenues for treating various diseases and infections.

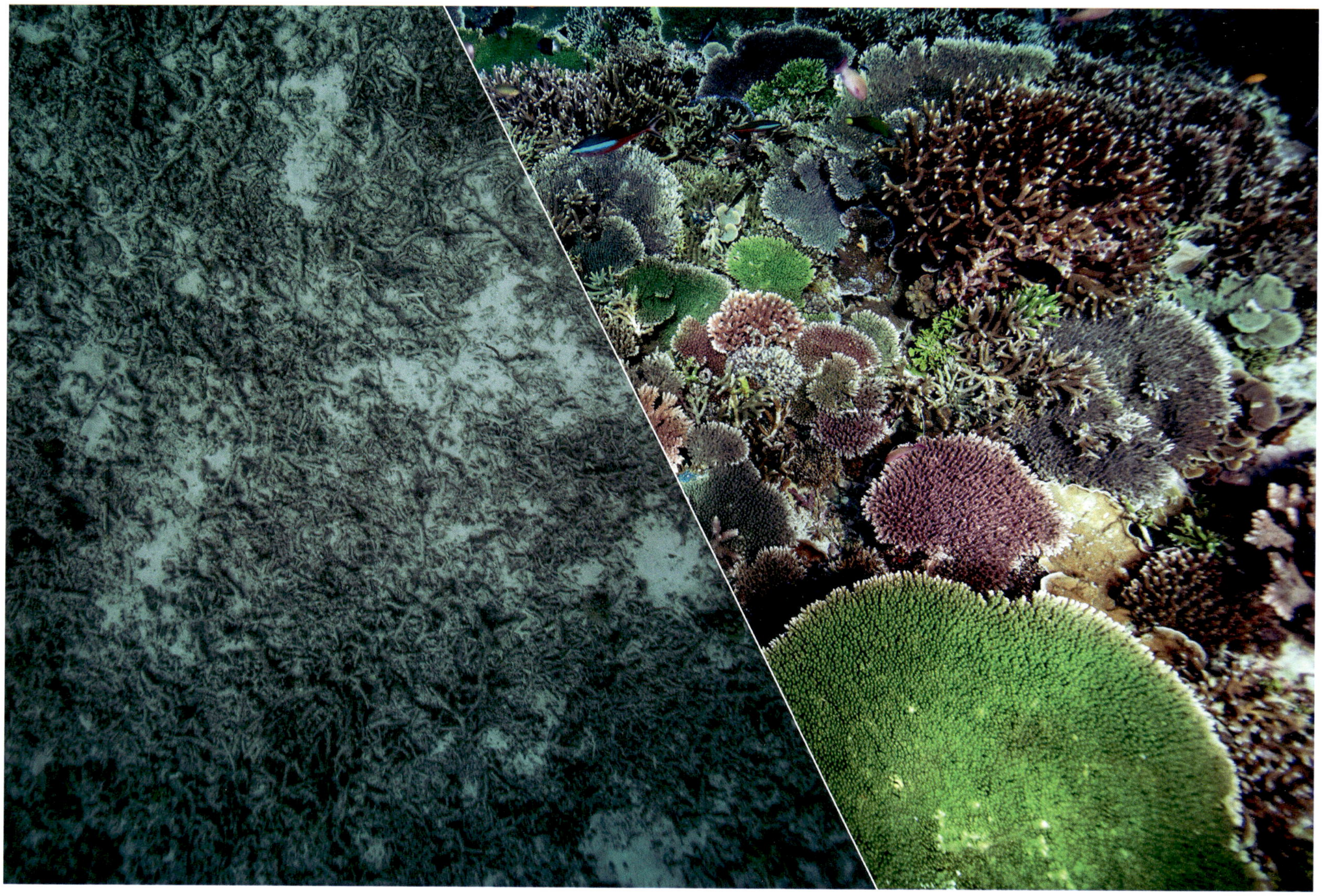

Coral reefs are in danger

Over the last 240 million years, coral reefs have evolved to become one of the most important and most complex ecosystems on the planet. But climate change and local dangers mean that they are at risk of dying out. If we don't act now, some scientists say that the world's coral could disappear completely by 2050.

A 2°C
increase in global temperature is the limit beyond which the world's coral would not survive.

40%
of coral reefs and mangroves have disappeared in the last 40 years.

ALMOST 90%
of coral reefs are at risk of dying out by 2050.

Coral Guardian

Founded in 2012, Coral Guardian is a French NGO that works internationally, supporting the conservation of coral ecosystems and the communities that depend on them.

Our mission is to involve local communities through the restoration, protection and development of coral ecosystems that have been damaged by human activity. Their development will help to create alternative sources of sustainable revenue.

Coral Guardian is an adventure that would never have been possible without the tireless dedication of our Indonesian and French team. We are also fortunate to have the daily support of people around the world. We would like to thank them for their unwavering commitment to the cause.

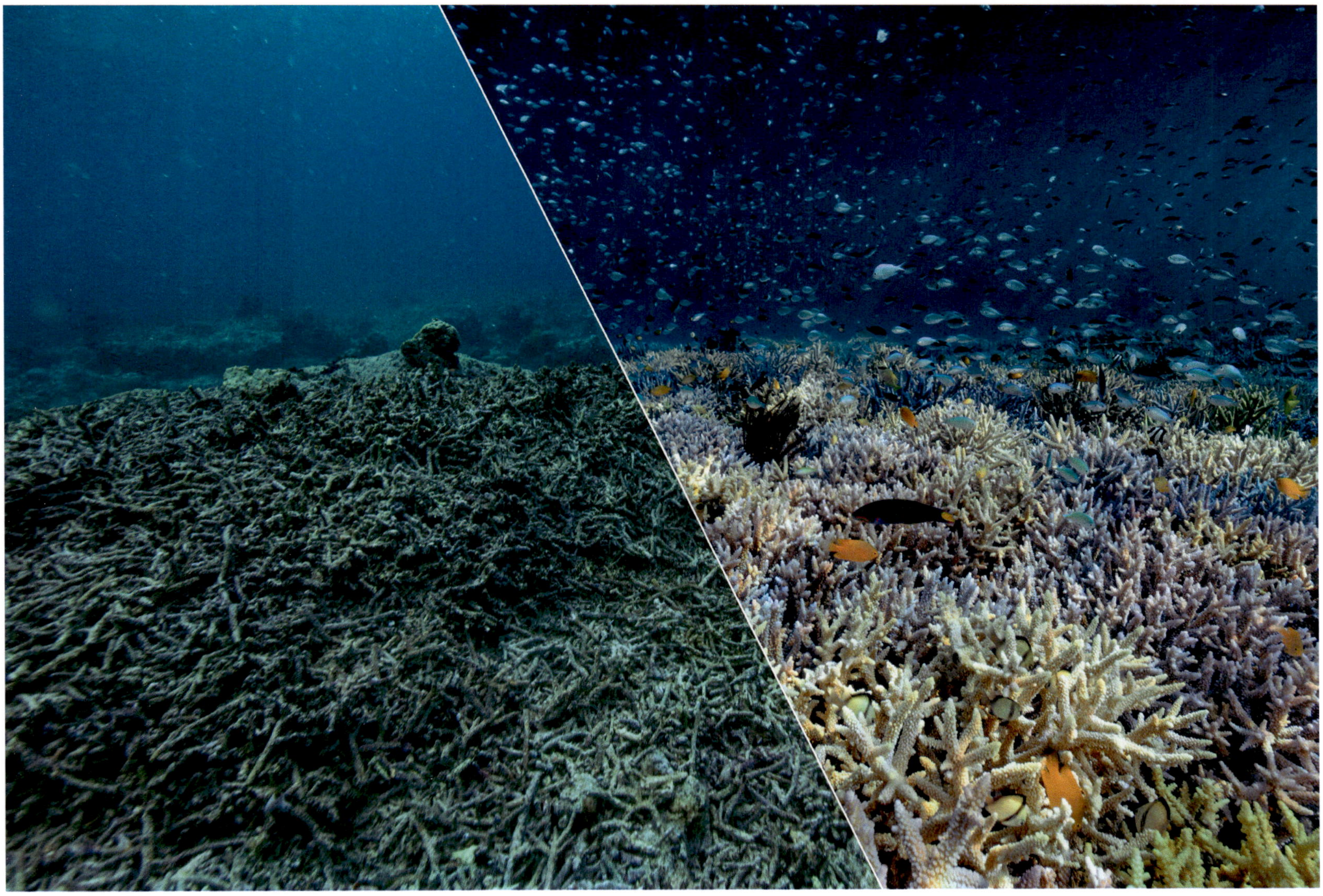

Our mission

Our organisation functions according to three main principles: participative marine conservation, raising awareness and science.

PARTICIPATIVE MARINE CONSERVATION
Coral Guardian restores damaged coral reefs through coral fragmentation and close scientific monitoring. The reef's restoration is a means to an end, not the end goal. It enables local fishermen to better understand their natural environment in order to protect it in the future. It is clear that we can't turn back time, but we can work with the current situation and make the most of it.

RAISING AWARENESS
Coral restoration is also a tool for raising awareness. Coral Guardian teaches communities about their natural environment on a local level by promoting sustainable fishing practices and on an international level through its 'Adopt a coral' programme, its raising awareness kit and a number of events.

SCIENCE
Coral Guardian studies scientific research linked to different areas of restoration, particularly how coral can adapt to global warming. Our team is constantly innovating in order to improve our techniques and try to maximise the growth rate of transplanted coral and the return of biodiversity.

PROMOTING OUR MODEL ACROSS THE WORLD
The organisation usually gets the same question from all over the world: how can we combine environmental and social issues in the same project? In response to that, we've decided to develop theory-based and practical training courses through 'The Blue Center'.
The Blue Center underlines our participative way of working: members of local communities who have been trained in coral aquaculture then become trainers themselves. It adds value to their experience and benefits others looking to get involved as well. The Blue Center will allow us to develop a network and a movement of people around the world who are engaged in the protection of coral reefs.

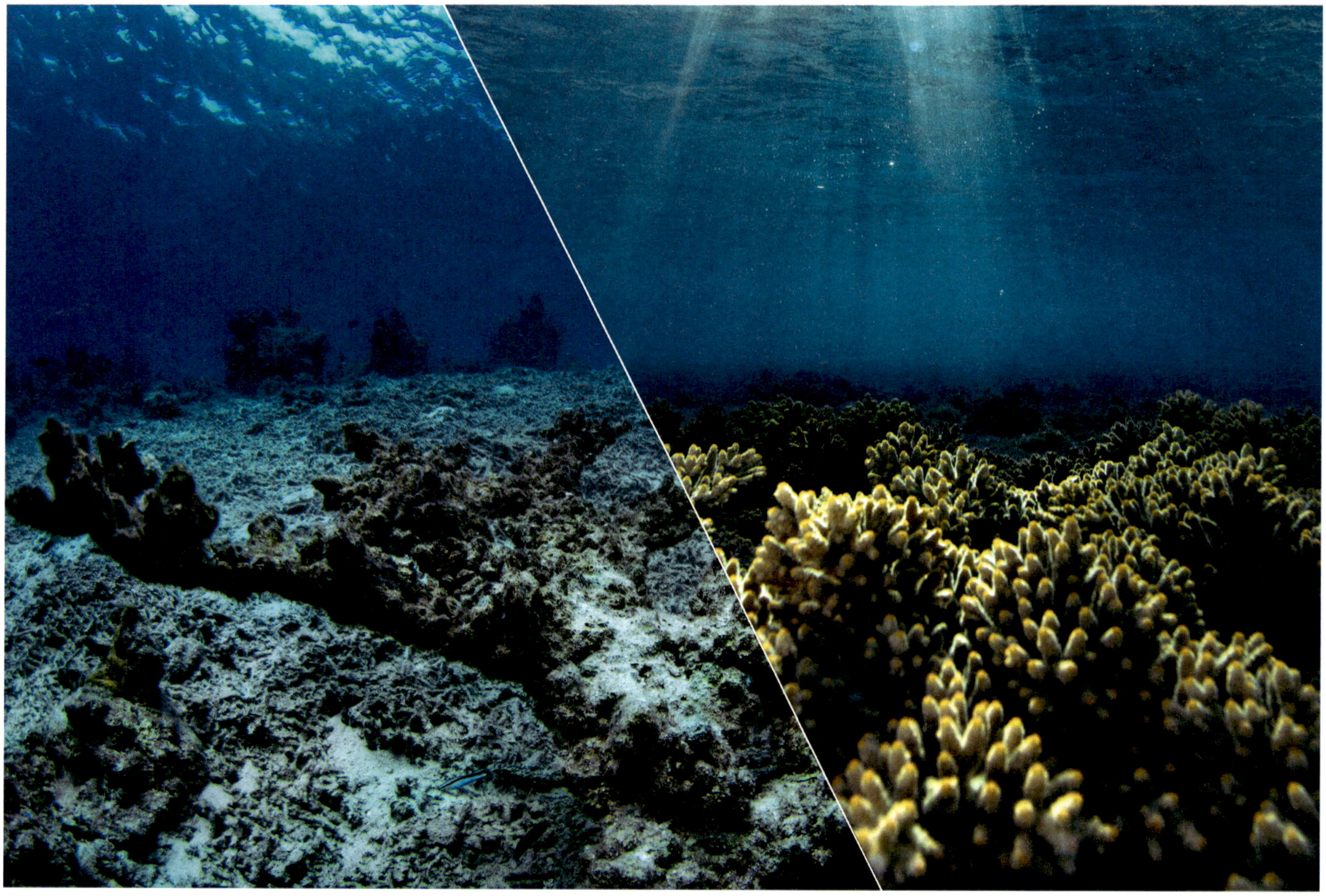

A PHOTO REPORT ON OUR PILOT PROJECT ON HATAMIN ISLAND IN INDONESIA

Since 2015, our organisation has been working on Hatamin Island in collaboration with the local NGO WES. This area to the north-west of Flores, near Komodo National Park, was designated a marine protected area in September 2019 by the Indonesian government at our request.

Overfishing and blast fishing have affected the balance of local coral ecosystems and disturbed the whole food chain.

A staff of 8 people, who are predominantly former fishermen, work full time on the project to restore and protect coral reefs in the area. Each month follows a work schedule that involves restoration, raising awareness, scientific monitoring and protecting the marine protected area.

A social monitoring group has also been established with the aim of helping to managing local marine resources.
Every month, 10 fishermen are interviewed about their fishing habits.

As for the environmental monitoring of the transplanted coral, the local team keeps a close eye on how the transplanted reefs are growing, observing numerous indicators such as growth rate, mortality rate, bleaching rate and the return of biodiversity.

This will result in the improvement of local communities' quality of life thanks to the regeneration of fish stocks.

KEY FIGURES

40,000 corals transplanted in the last 5 years
30.2 times more fish in 4 years
1.2 hectares of marine protected area created
30 local jobs created since the beginning of the project

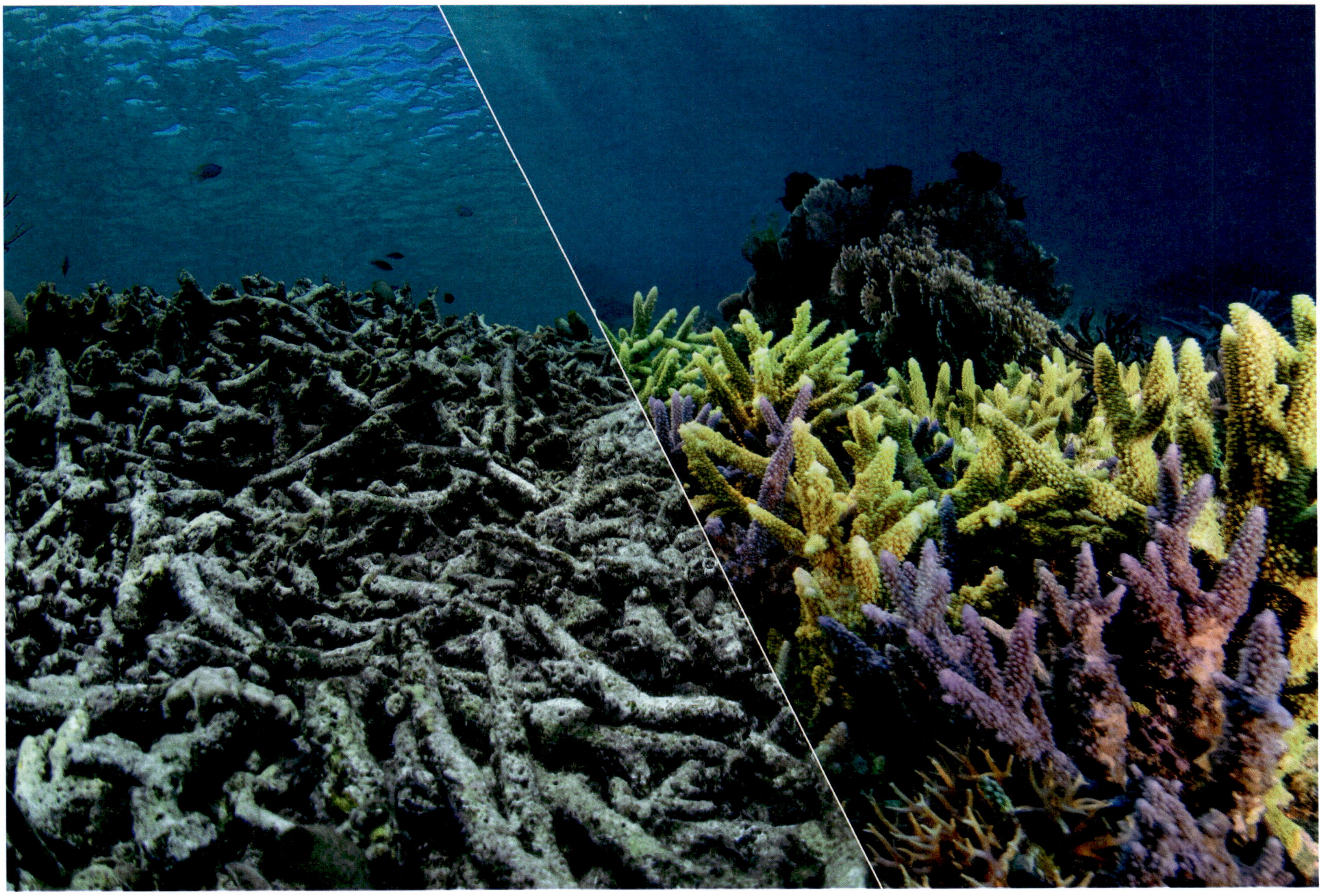

The comeback. This is the progress made in only three years in a restored and protected zone. It's the perfect example of the return of a coral reef in a zone where the seabed is constantly moving.
These structures we set up offer solid support to transplanted reefs. The environmental conditions are perfect for bringing life back to the area.

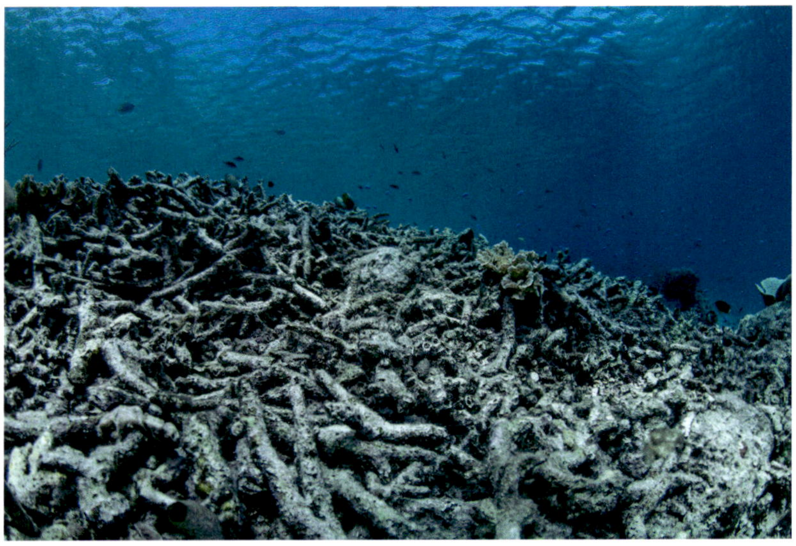

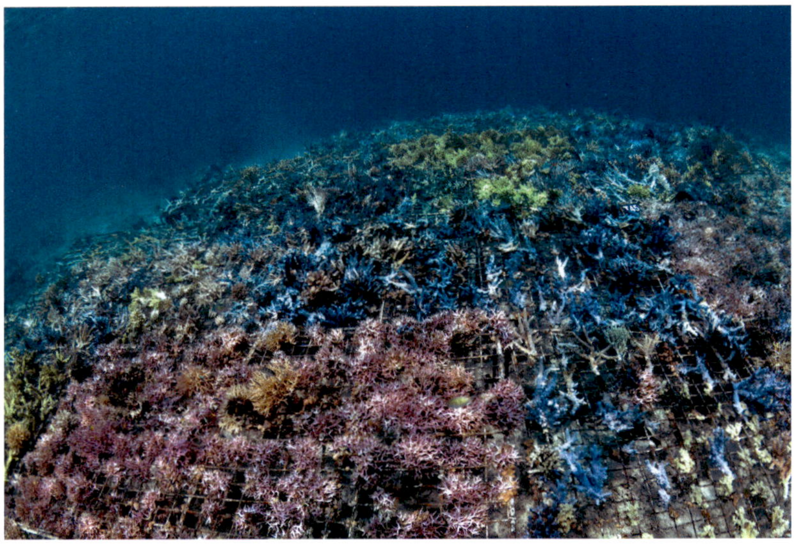

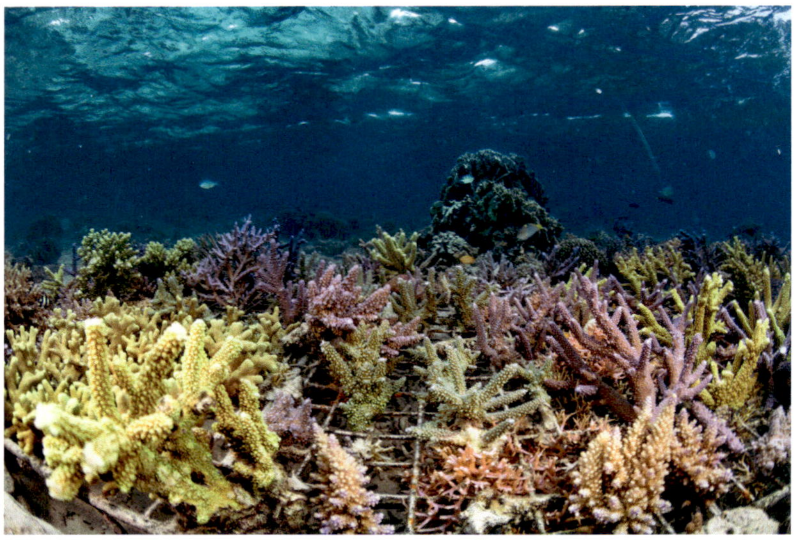

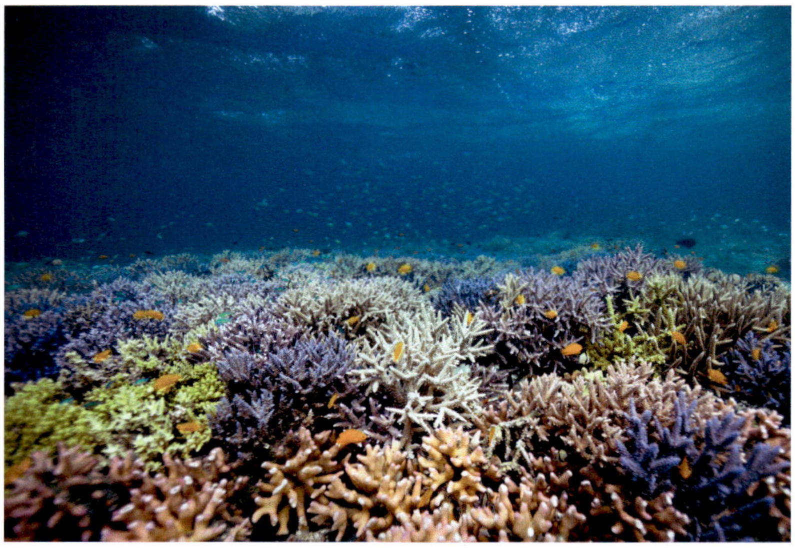

Transplantation. The team is in the process of carrying out a transplantation. The coral colonies are fragmented and then solidly attached to a metal frame. A month later, thanks to their limestone skeleton, these corals will have enveloped the structure to become more stable and then will start to grow towards the sun.

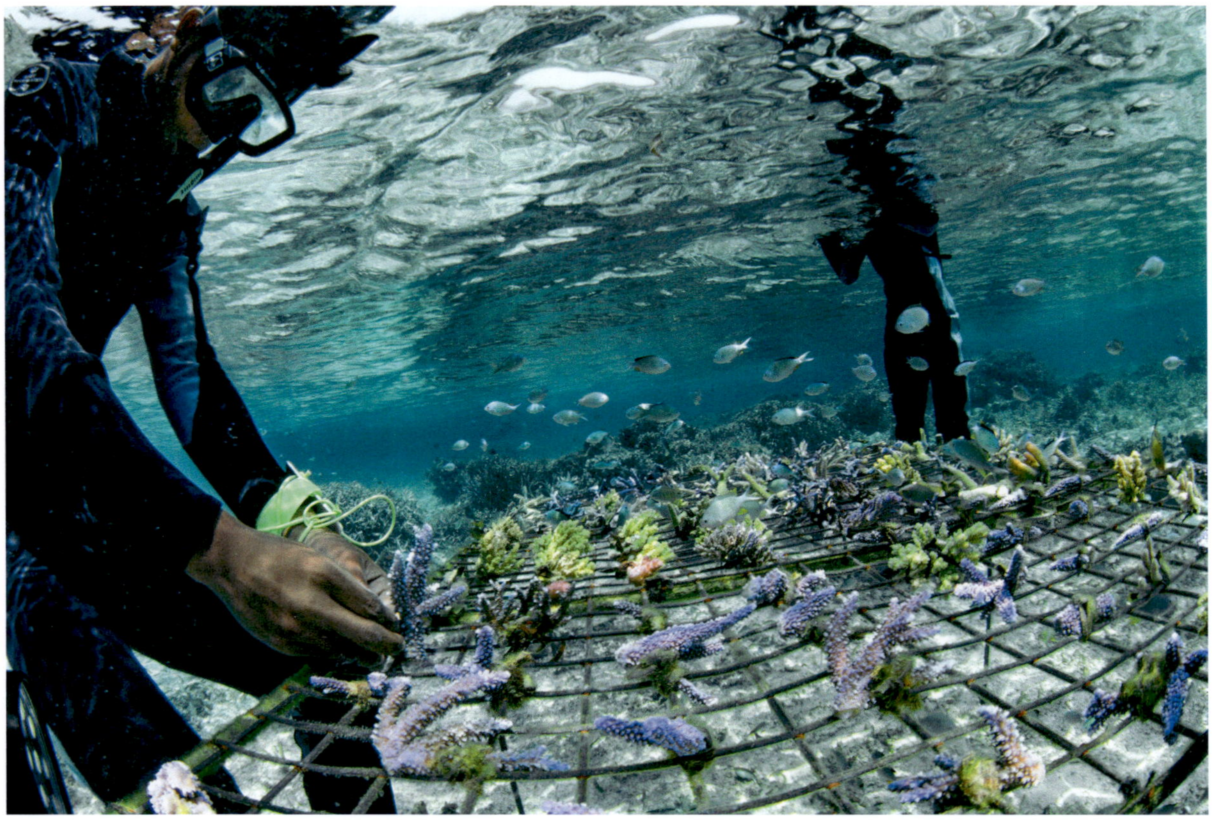

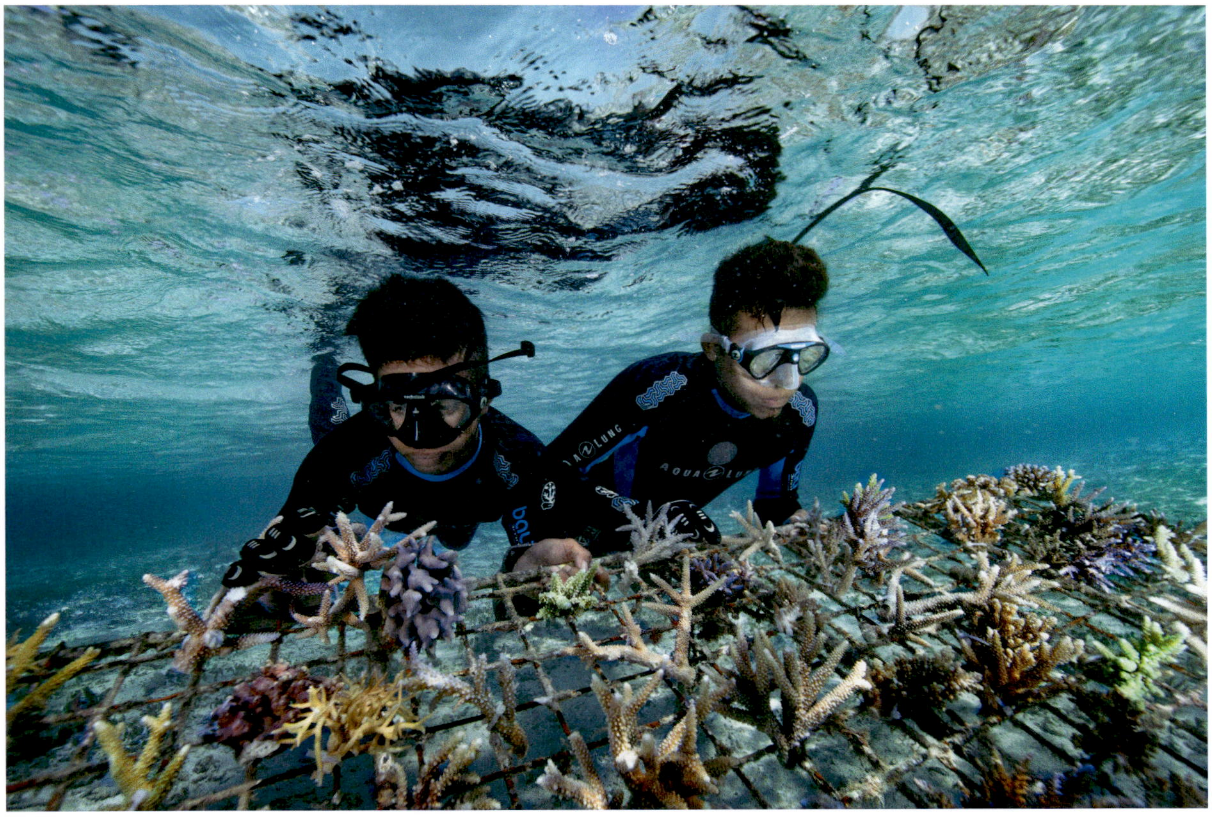

Harvest time. When the coral colonies grow on the frame and reach an adequate height, it's possible to harvest a bit to create new fragments for transplantation. It's a good alternative to harvesting wild coral.

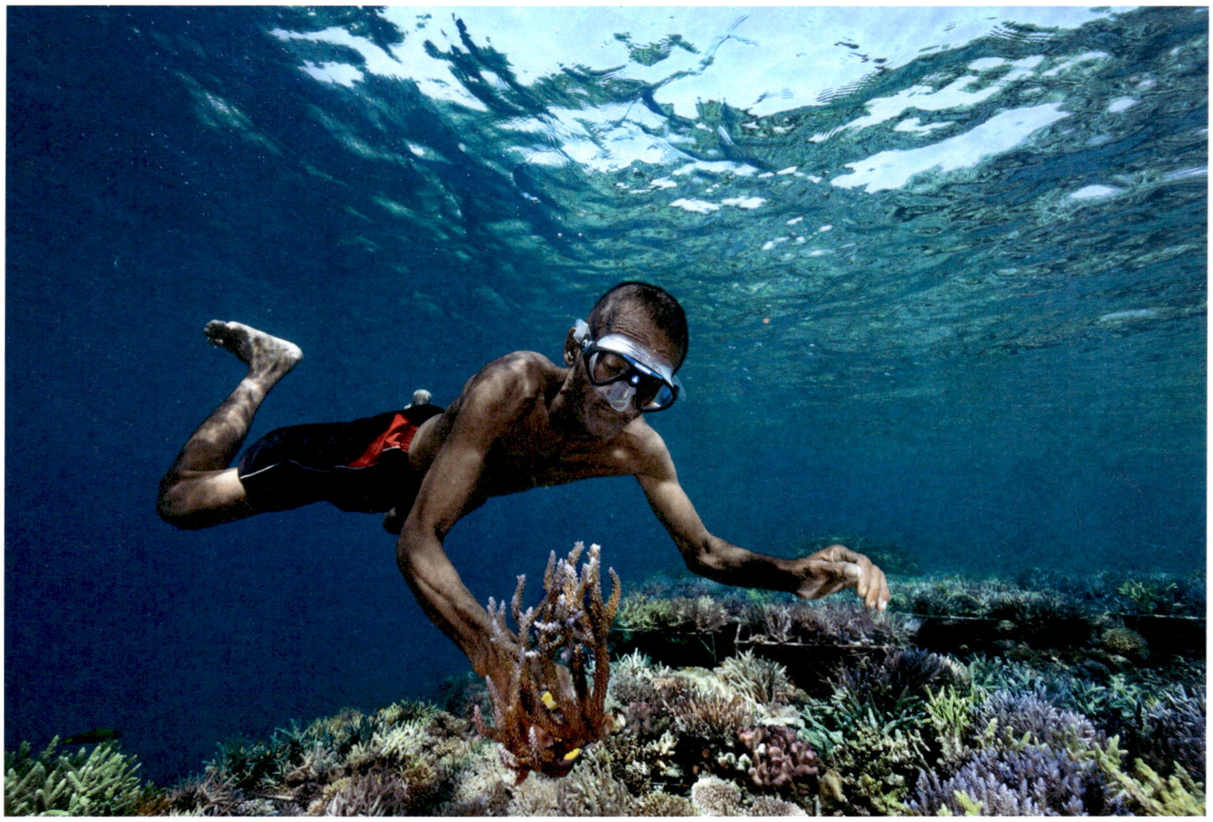

Concentration. The transplantation of coral colonies has to be done underwater to avoid causing stress for the coral above the surface. To make the task easier, the coral is attached to structures that aren't very deep underwater before being transported to the final site where they can grow in optimal conditions.

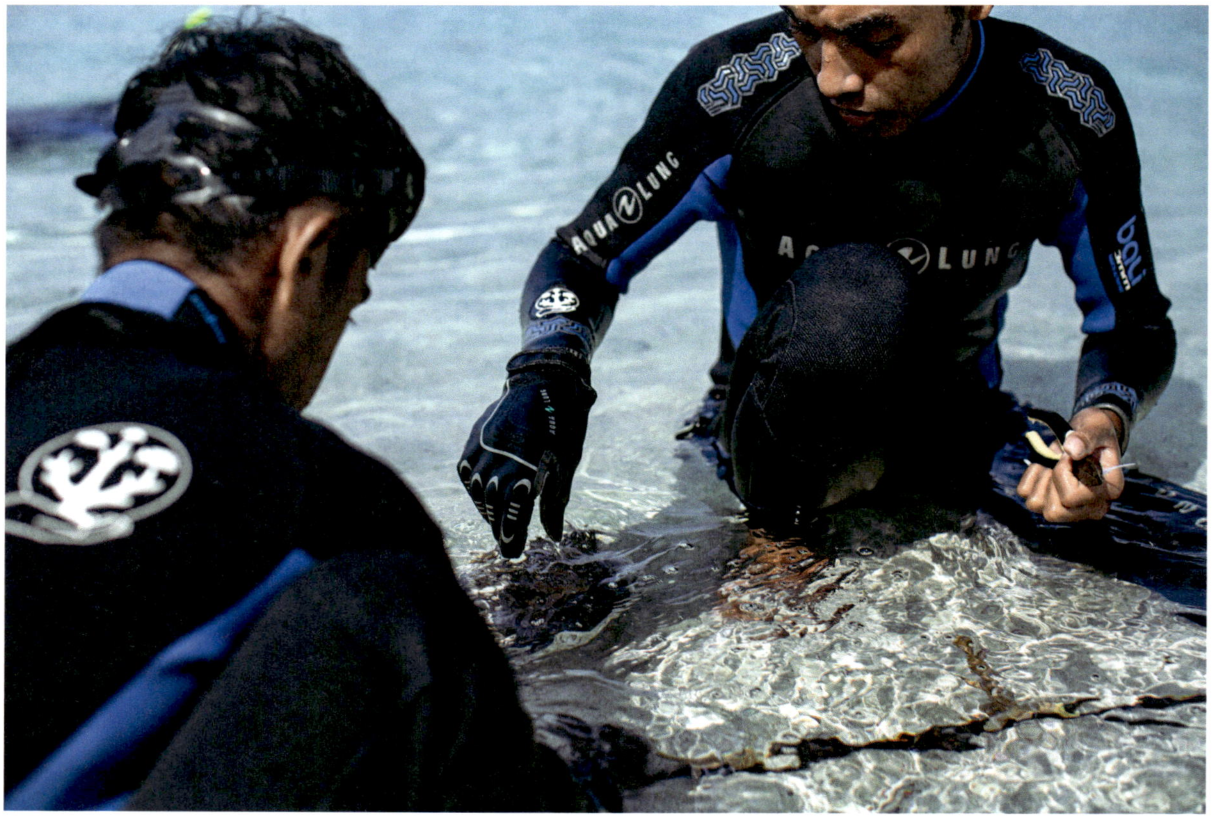

Predator. This is the *Acanthaster planci*, a large starfish that consumes coral. The colony you can see on the left has already fallen victim to it. The starfish will leave behind white skeletons sucked dry of all nutrients and substance. The number of these starfish is increasing every year because of a lack of predators. This could become a real problem for the reefs if their population becomes too large.

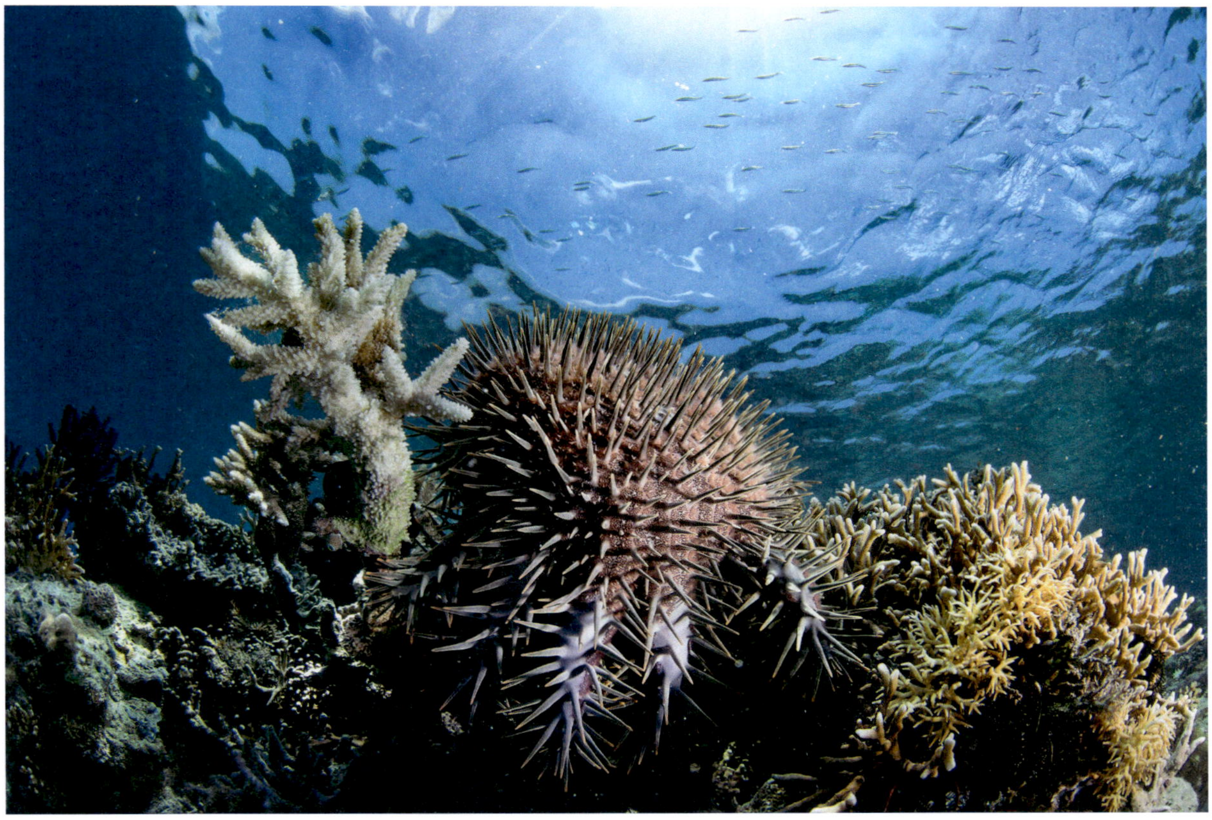

Resurrection. A restored coral reef that is only three years old. Its colonies here are mostly made up of the *Acropora* coral. The growth rate of certain species can be up to a centimetre a month. Before starting the restoration process, this area was deserted of sea life. The results are astonishing.

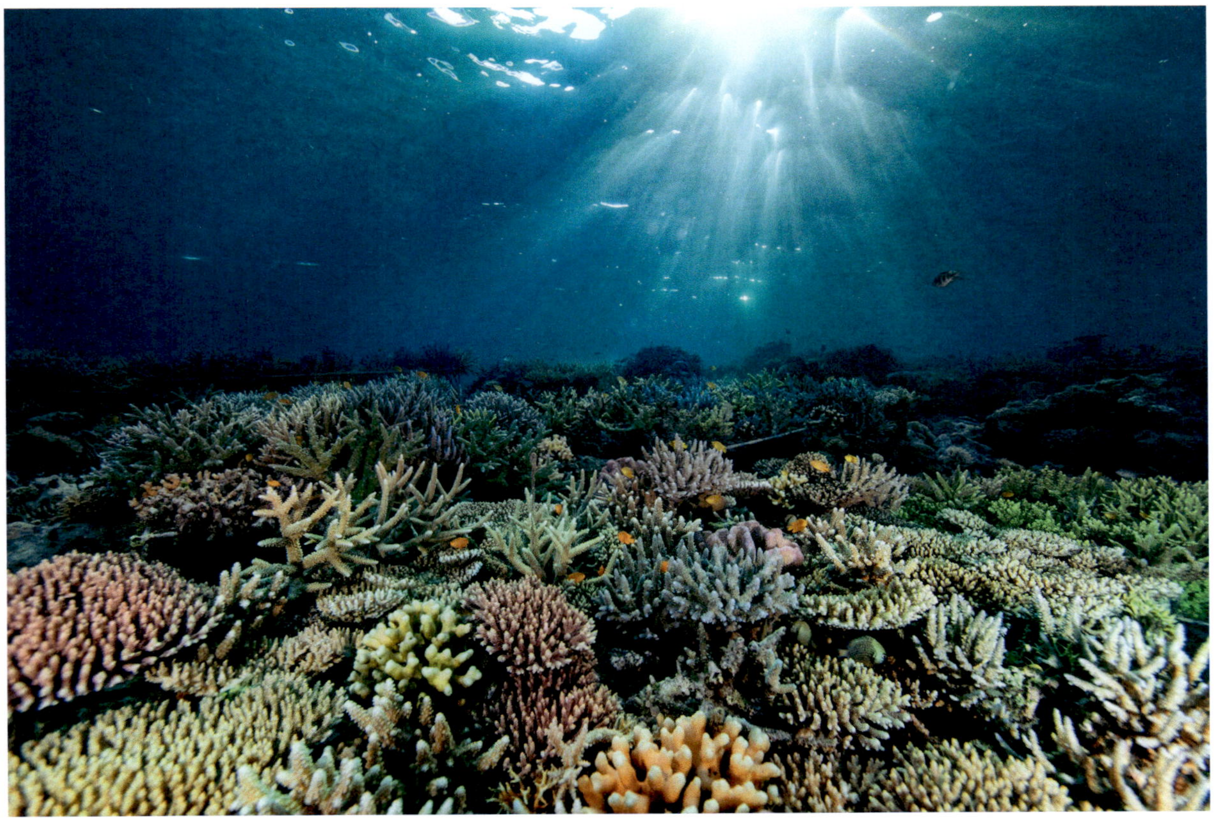

Science. Scientific surveys are essential to understand and monitor the growth of a restored and protected zone over several years. A dive mission involves counting the different species of fish as well as their abundance by using a line laid down at the bottom of the water. This process is repeated once a year at the same time.

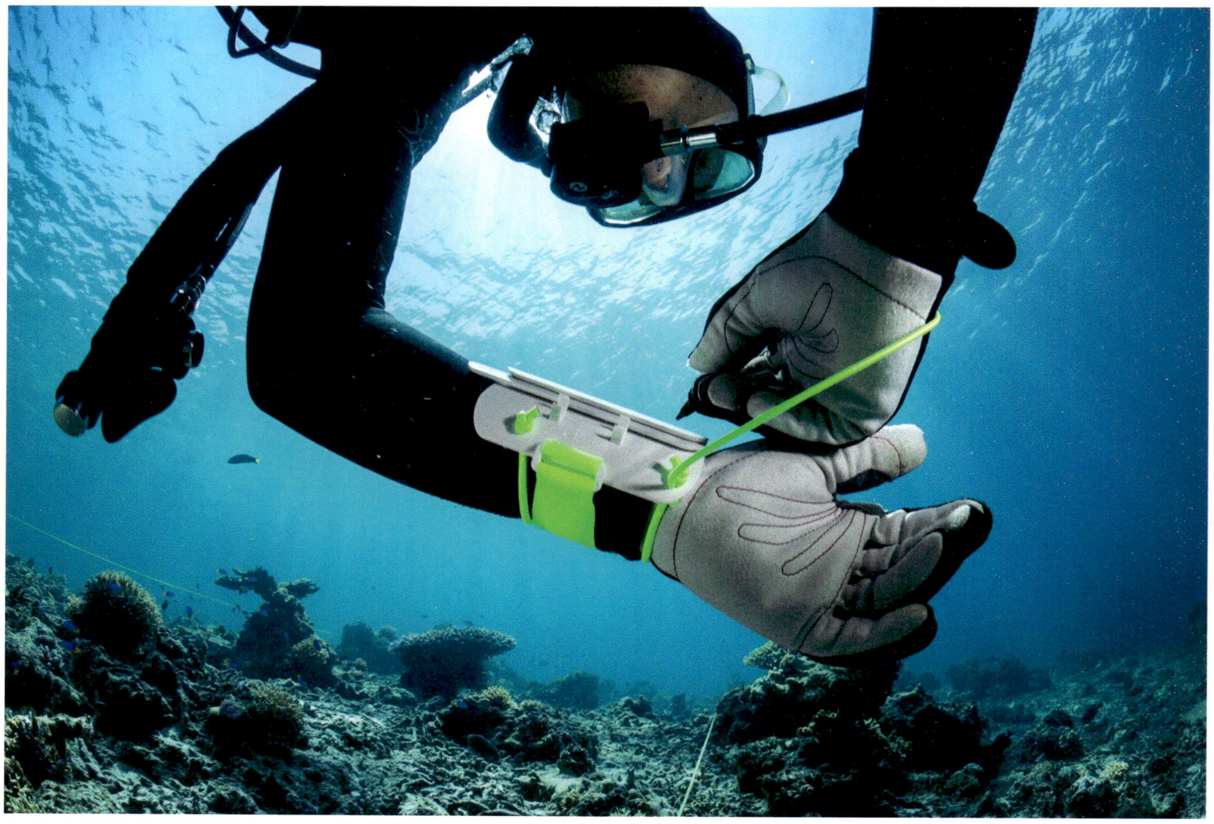

Swimming in paradise. This reef is only three years old and is made up of a variety of coral species (*Pocillopora*, *Stylophora*, *Acropora*, *Seriatopora*). The speed of their growth is remarkable when you consider that they were only a few centimetres long when they were transplanted.

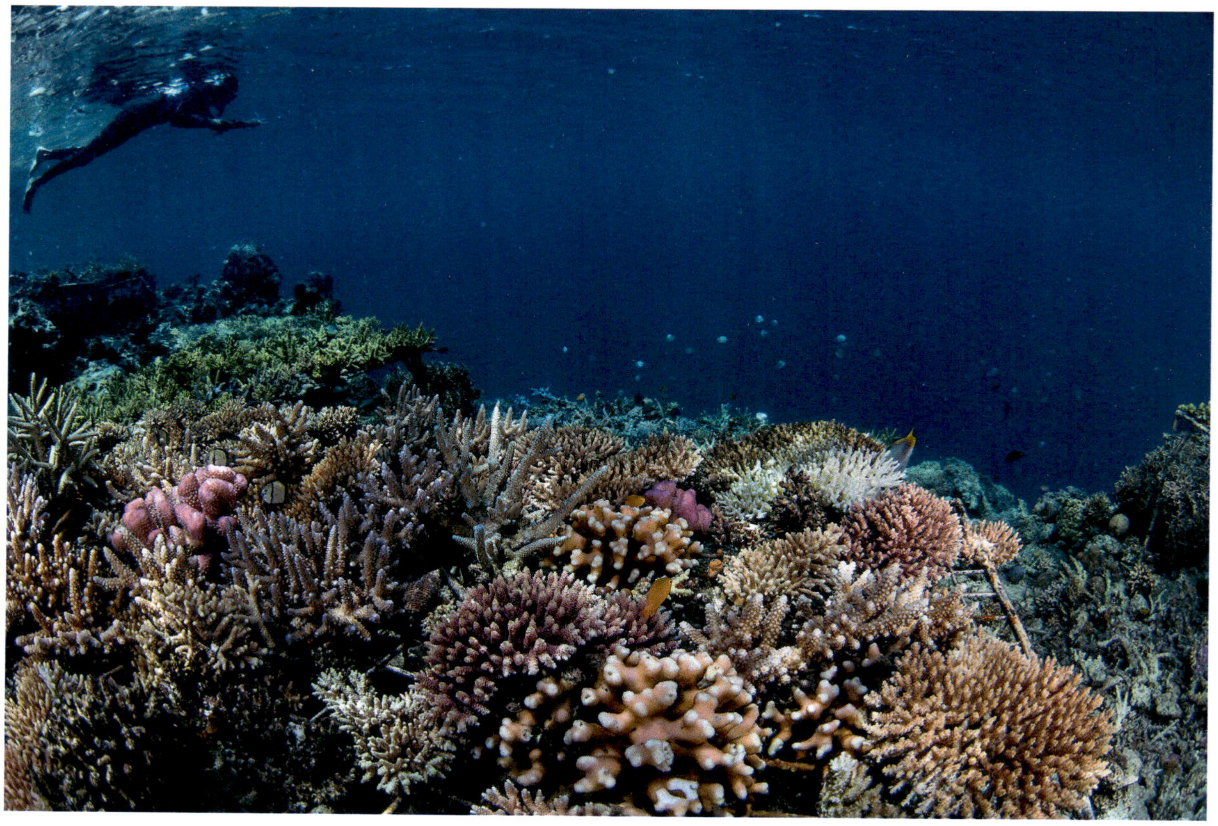

Monitoring. Monitoring and maintaining the reef is every-day work. We have to regularly test the coral's general health, any potential parasites present, collect broken coral around the structures and transplant them elsewhere. In this photo, Sahril has just picked up a broken coral from the seabed and is transplanting it into a nearby structure.

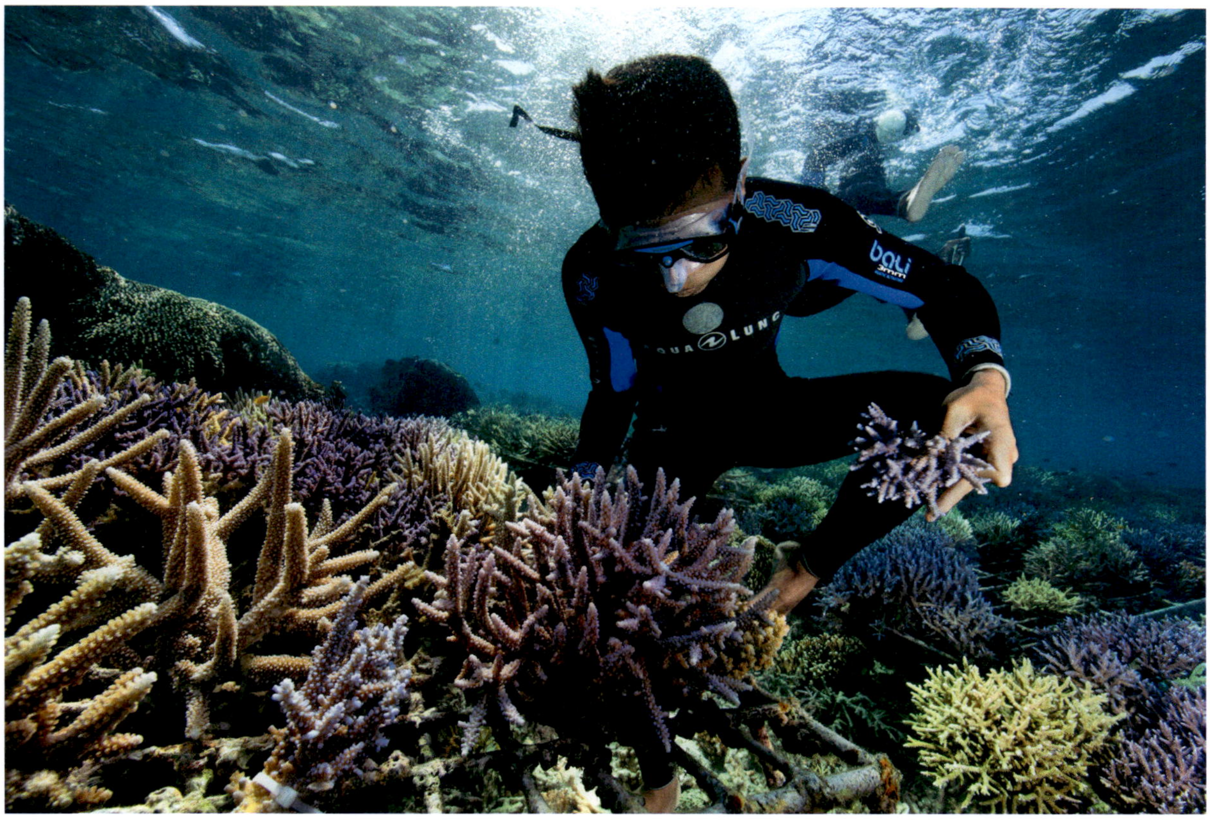

Maintenance. Maintaining all of the equipment is crucial so that everything functions correctly. The buoy line that marks the perimeter of the marine protected area is cleaned several times a year.

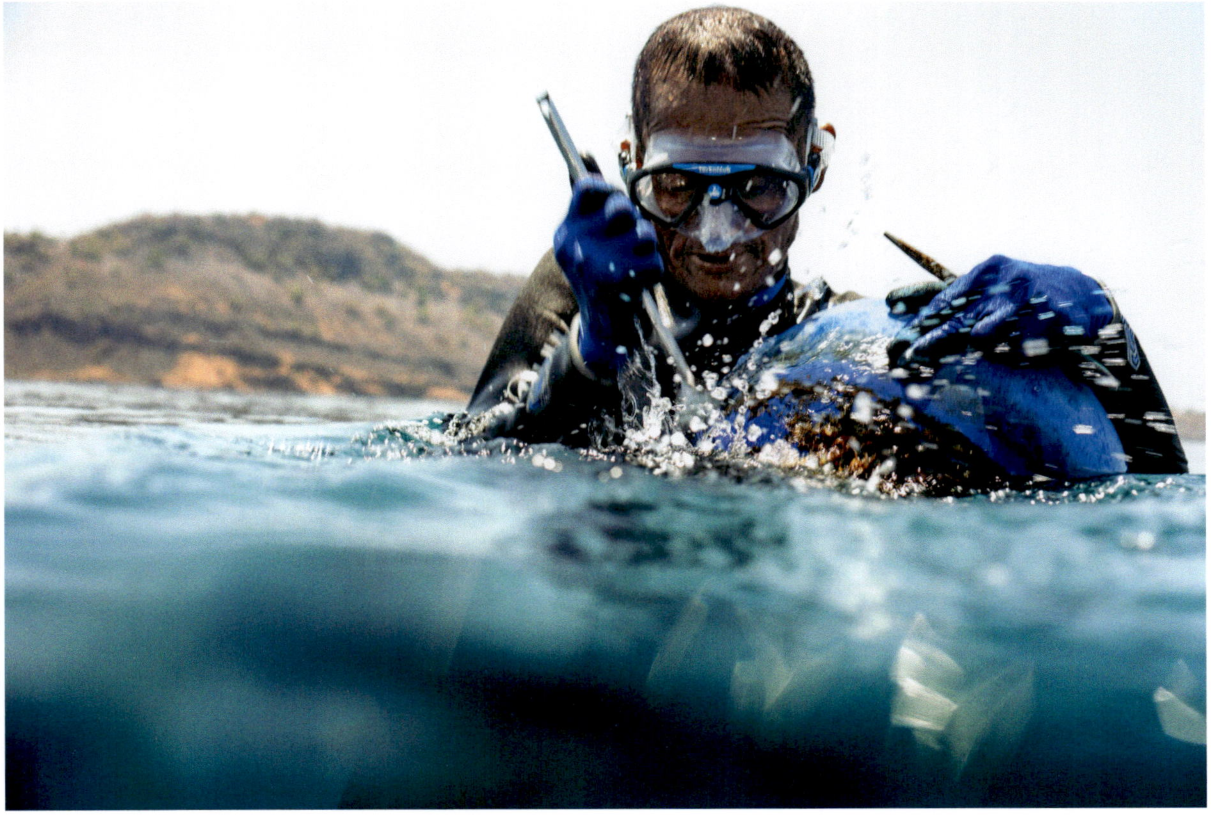

Acknowledgements

First of all, I'd like to thank all of the residents of the village Seraya Besar, for whom I have the deepest admiration and respect. They trusted me enough to include me in their daily lives and let me participate in their ancestral traditions such as fishing. I am immensely grateful to them for the rich experiences we shared. I would like to thank them for welcoming me, for their friendship and for sharing with me their zest for life. I am aware of just how valuable this experience was.

Coral would never have been possible without the indefatigable work ethic of the Indonesian and French teams behind the NGO Coral Guardian. I extend to them my warmest thanks. I would like to thank Julien Holleville in particular, who has been a colleague since the organisation's creation but a friend for much longer, and who provided precious support during the most difficult times.

I would also like to express my gratitude to the publishing house Hemeria, to Amélie Boutry as well as Valérie Van Oost, who believed in this photography project and introduced me to the world of publishing and all of its quirks.

I would also like to take a moment to acknowledge the businesses and foundations who are partners of the project as well as all of the people who pre-ordered *Coral*: H2O at Home, fondation Lemarchand, Fondation WES and Coral Guardian.

Finally, this book could not have seen the light of day without all the people who pre-ordered the *Coral* book, thus demonstrating their confidence.

From the bottom of my heart, a huge THANK YOU to everyone for your support!

The contributors: Raphaël Abdou (Team Session), Frédéric Agnès (Compilatio SAS), Jessica Algrain, Aldine Amiel, Virginie Amilien, Maryline Andrieux, Luciana Aquino, Valentin Arnaud Tarenque (Les fruits de Roquelongue), Albert Arock, Alice Arthaud, Xavier Assemat, Armelle Assouad, Clarisse Auffret, Étienne Aygalenq, Thomas et Pauline Bachellerie-Colognoli, Emmanuel Bapt, Grégoire Bastide (Escale Création / Luxaqua), Georges Baudvin (Make It - L'Atelier Maquettes), Jules Baudvin (Lug Productions), Guy Bellon, Patricia Benoit de Nyvenheim, Maxime Beraud, Caroline Bernard (Module), Danielle Bernard, Romain Bernard, Maxime & Lucie Bernardi, Colette Berruyer, Cécile Berthault, Mag & Aurel Bertini / Jeanningros, Thibault Bescou, Martine Besgnard, Christine Bessard, Hervé & Edwige Bessi (S.M.B.), Martin Besson, Bénédicte Billaud, Carole Billoud, Thibaut Blind, Vincent Bocquee, Yoan Bonhomme, Guénola Bonilla, Fabrice Bonnaventure, Julie Borgeot, Audrey Borras, Evelyne Bouancheau, Stéphane Boulant, Anaïs Boungnasiri, Julia Bource, Caroline Bourgeois, Isabelle Bousquet, Florian Braun, Bruno Breton, Camille Breton, Hervé Briandet (Clinique Sud Vendée), Arnaud Brival (Association The Sea People), Florent Brozzoni, Patrice Bureau (Longitude 181), Romain Burret, Nicolas Camus, Miguel Candelas, Alain Carpentier, Jean-Paul Carrot, Cyrielle Cayrol, Pascale Céron, Ludovic Chalamel, Jean-Pierre Chaligné, Coralie Chambron, Guillaume Chazot, Mikaël Chevalley Coola Corporation, Emmanuel Chollet, Laura Crisan, Delphine Christ, Julien Colognoli, Pauline Colognoli, Robert Colognoli, Bénédicte Colognoli Pey, Jérôme Comin Medallia, Corinne Copin, Aurelien Corfu, Youri Courrege, Hélène Coutard, Stéphane Crémier, Olivier Cros (Meggitt Sensorex), Marta Csabi, Sandra Cuello, Françoise Cuny, Julien Dagois, Adélie Daniere (Make It), Véronique Danois, Solene Dargaud, Béatrice Datrino, Laurence Daubignard, Marion Daubignard, Sylvie Daubignard, Florence de Ferran, Isabelle de La Peine, Cathy Debailleul, Laura Degryse, Elise Delandre, Charles Denis Charlesdenis, Martin Desmalades, Olivier Detournay (Planktovie), Jean-Luc Diainville, Anne-France Didier, Arvanitidou Dimitra, Monsieur Dioni (Chupalo records), Aurélie Dubois (Association RespectOcean), Olivier Duguet, Dominique Dumaine, Clémentine Dupuy-Roudel, Julien Dussauge, Bruce Eernisse (Banggai Consulting Ltd), Diane Enkelmann (Coralaxy GmbH), Katie Erbacher, José Escaño Roepstorff, Olivier Eude, Latour Fabrice, Valérie Fahrner, Adrien Fargier, Stéphanie Fellen (FLN ADVISORY SPRL), Éric Fellous, Nastazia Femmami, Antoine Fenaux, Émilie Ferrer, Lucie Ferrer, Coralie Ferrier, Franck Fougère, Nicolas Fraisse, Frédéric Francois-Bongarcon, Pascale Fremeaux, Pierre-Jean Fréville, Murielle Froger, Karine Furaut, François Gaillard, Maël Galland, Jérémy Gallemard, Jean-Pierre Galot (Santé au travail du Genevois), Anne Garcia, Olivia Garcia, Laura Garrido, Marie Gaspard, Erwan Gensac, Paul Gerbeaud (The Sea People), Yoann Gerboud (G. Architecture), Serge Gil, Myriam Goisque, Alain Golea, Joëlle Gourier, Luc Gravel, Bastide Gregoire (Escale Création / Luxaqua), Elisa Grillet, Grégory & Magali Grossin-Guéant, Éric Guenard, Jackie Guillermand, Zelie Haeuw, Marine Hamelin, Annie Hamm, Cécile Hamm, Sören Hänsler, Niels Heikes, Patricia Hivert, Jean Houze de L'Aulnoit et Mathieu Colin (Limestreams), Clément Hubrecht, Nicole Huez, Yvan Hunault Buceo Almuñecar, Bryan Jacquemard, Gilles Jaillard, Clémentine Jeremy, Lesley-Ann Johnston, Sylvie Jouve, Simon Julien (huîtres-bouzigues.com), Millot Juliette, Nicolas Jullien, Éric Kaiser, Eduardo Koga, Katalin Krajnyak, Katalin Krajnyak, Astrid Kunst, Sophie Labrut, Floriane Lafore, Justine Laurent Wiithaa / Circulab, Luc Laurent (BioInsight), Édouard Le Gall, Le Reboot des Mères lyonnaises, Phillippe Lesur, Hervé Lobet, Romain Loison, Camille Long, Hélène Lotte, Patrick Louvet, Justine Luqué, Andrée Maes, Estelle Maître, Timothy Man, Geoffroy Maquet, Ghislaine Marigny, Charlotte Martin, Isabelle Mary, Laurence Marzin, Julie Masegosa, Patricia Massé, Guillaume Masson, Jérôme Mathey (DroneGo), Lucile Mazodier, Carine Meerschman, Sarah Mension, Sylvain Mestre, Jean Mezard, Thomas Miard, Julian Miribel, Éric Mondon, Pablo Montoto, Christine Moran, Florian Morel, Camille Morel, Isabelle Morelon, Marie-Noëlle Motroni, Anne-Sophie Mouraud, Eleni Mouroutsou, Serge Muller (Plongeurs du Monde), Philippe Mussi, Jean-Philippe Nicolay, Laura Nombret, Jean-Laurent (Ocera Areco électricité), Pierre-Antoine Ollinger, Solène Ollivier, Grégory Pagliero, Rosemarie Pailhes, Marina Palacios Particular, Guillaume Papoz, Lise Pauly, Valentine Payre, Éric Perrier, Nathalie Perron, Christine Pey, Louis Pey, Noémie Pey, Nicolas Philippe, Rémi Pillot, Stéphane Piton, Muriel Piwowarczyk, Geneviève Poirot, Jano Pontarollo, Cyrille Poujol, François Poujol de Molliens, Coralie Probst (F2M), Guillaume Pujol (Marinehome), Fabien Rambert (CREAFA), Michel Ramel, Vijay Ramnauth, Olivier Raybaud, Marie Reber, Bruno Rebout, Valérie Remize, Paul-Étienne Remy, Nicolas Richard, Benjamin Rico, Coline Rigole, Stéphane Rochon, Olivares Romain, Alexis Roman, Jeanine Rondeaux, Léon Roures-Henriot (Léon Dsgn), Christophe Rousseau, Lionel Roux (SCI mathisimon), Fanny Rubia, Jean Ruff, Marc Sadin, Pascale Salaün (BiocenOSE Consulting), Ève Seguy, Nicolas Selosse, Daniel Serant, Jean-Claude Serre, Krista Shoe (Mother of Corals), Wendy Skinner Buet, Maeva Soubrillard, Philippe Spadotto, Mélodie Sperandio, Sarah Stiffel, Thibaut Stremplewski, Florence Tabourel, François-Xavier Tainturier, Yassine Tajja (RTE), Alban Taravello (W2P PRODUCTION), Frédéric Tauzin, Xavier Tétard, Busson Téva, Lorin Thorax, Taï Trinh Reef Technics, Ralph Unger, Joe Unsworth, Stefanie Van Zyl, Jean-Dominique Vanpe, Isabelle Verdier, Julie Vermersch, Jean-Louis Verrière, Cédric Vetillart (SAS Traversée / Mandara), Blandine Viadere, Michel Viallon, Noémie Vilain Perini, Géraldine Villani, Robert Villani, Maran Vincent (Commission Biologie FFESSM), Jane Williamson, Lena Zabarino, Philippe Zabarino, Éric Zabukovec, Alexandra Zivy.

Photographs Martin Colognoli

Texts Valérie Van Oost
Copy Editing Fabienne Texier
Translation Catherine Bennett
Graphic design and layout Amélie Boutry
Image processing and printing printmodel®, Paris

ISBN: 978-2-490952-32-8

Printed in November 2021 in France
Legal Deposit January 2022

@hemeriaphoto, Paris
#ExceptionalPhotoBooks
www.hemeria.com

In support of French organization ReforestAction,
for each copy of this book purchased on
www.hemeria.com a tree will be planted.

reforestACTION

HEMERIA